THE COMPLETE HANDBOOK OF LAWNMOWER REPAIR

DEDICATION

For my beloved parents

No. 767
$9.95

THE COMPLETE HANDBOOK OF LAWNMOWER REPAIR

BY PAUL DEMPSEY

TAB BOOKS
Blue Ridge Summit, Pa. 17214

FIRST EDITION

FIRST PRINTING—APRIL 1975
SECOND PRINTING—NOVEMBER 1975
THIRD PRINTING—MARCH 1976

Copyright © 1975 by TAB BOOKS

Printed in the United States
of America

Hardbound Edition: International Standard Book No. 0-8306-5767-3

Paperbound Edition: International Standard Book No. 0-8306-4767-8

Library of Congress Card Number: 74-33622

Preface

This book is intended to be a complete guide to lawnmower repair; it covers all aspects of the art from the most elementary repairs to major engine and transmission work. The emphasis is on the more popular rotary and riding machines produced by Briggs & Stratton, Clinton, Kohler, Lauson, Lawnboy, and Power Products—whose engines power more than 90% of all lawn and garden equipment sold in this country. In addition there is a chapter on electric mowers, including the new cordless types; as far as we know, this is the first time that the principles of operation and repair for these machines have been made public. Another chapter is devoted to the Wankel engine, only recently adapted to lawnmowers.

Intended primarily for the owner who wishes to repair his own garden equipment, the book should also be valuable to the professional mechanic. For those whose interest goes deeper than repair, a long chapter on 2- and 4-cycle theory is provided.

The bulk of the material is based on factory manuals and factory training courses. However, it is more than a compendium; in addition to explaining how things work and why certain repair procedures are used, the author presents the material from the point of view of the working mechanic. You will find many shortcuts and "tricks of the trade" here; they have been gleaned from my own long apprenticeship in the mechanic's trade.

For their cooperation in furnishing much valuable information, I thank all of the firms which have so generously aided in the preparation of this book, with special acknowledgment to Kohler of Kohler.

Paul Dempsey

Contents

Chapter 1

Engine ABC's

Lawnmower engines are mechanically simple, but the theory which informs their design embraces the same body of mathematics and experimental knowledge which is used to build aircraft and automotive powerplants. A mechanic does not need to know the math, but he must have a good—almost intuitive—conception of the *whys* of engine design. Most mechanics never have the opportunity to go beyond the level of parts changer. A few of the really knowledgeable men develop an understanding of, and sympathy for, metal and engines to a degree that permits them to take part in the creative process of design. These are mechanics not content to merely repair an engine; they want to understand how it works and why it failed. This kind of understanding can only come with some knowledge of automotive theory in addition to experience. No book can bestow experience—you have to skin your own knuckles. But this chapter can at least supply the answers to your immediate questions and, serve as a springboard to further investigation.

An understanding of basic terminology is essential before discussing any technical topic. In Fig. 1-1 the *piston* is shown sliding up and down in a metal tube called the *cylinder*. The upper end of the cylinder houses the *spark plug*. This is the area *(combustion chamber)* where gasoline and air react explosively in the presence of a spark to release heat energy. The pressure of expanding gases drives the piston downward with a force and speed akin to that of a bullet leaving a rifle. This force is transmitted through a rod connecting the piston and crankshaft. The upper end of the rod pivots on the piston pin (or *wristpin*), and the lower (large) end mates to the *crankshaft journal*. This arrangement of parts causes the reciprocating action of the piston to be translated to rotary

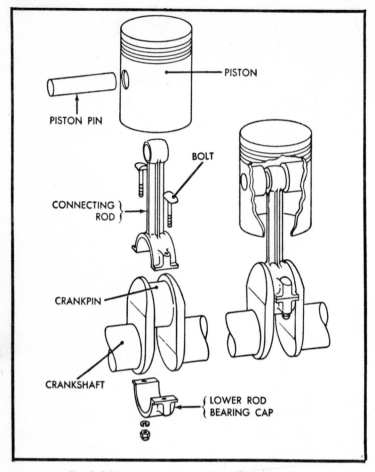

Fig. 1-1 Piston, connecting rod, and crankshaft.

motion in the crankshaft. Although not labeled in the drawing, the amount of crankshaft offset is called the *throw*; it determines how far the piston will travel in the cylinder. All full-size engine designs today use one or more sealing rings above the wristpin *boss* to prevent leakage of the burning gases; very early gasoline engines sometimes employed a piston without rings.

Of course, a working engine needs more in the way of hardware than a cylinder, piston, rod, and crankshaft. The engine must be "fed" a fuel/air mixture either by way of poppet valves or from ports machined into the sides of the cylinder. Valves are opened by a *camshaft* which is driven

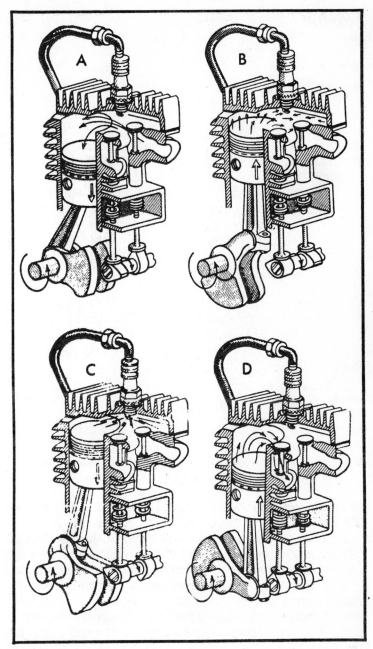

Fig. 1-2 Four-stroke-cycle in a Tecumseh side-valve engine: A—intake;
B—compression; C—power; D—exhaust.

from the crank through a pair of timing gears, and closed by springs. To keep things moving, a flywheel is mounted on the crankshaft; its momentum acts like a kind of energy bank—during the piston's downstroke it stores energy, then releases it on demand.

In addition, an engine needs a *carburetor* to mix gasoline with air in the right proportions and a spark generator. Some provision for cooling and lubrication is necessary. And since most of us have neighbors, we also need a muffler to reduce some of the explosive sounds escaping out the exhaust.

With the exception of electric motors, the engines we use to power lawn and garden equipment are in the class of internal-combustion powerplants. The energy used to drive the piston is developed internally, in the combustion chamber. In contrast, steam and the newly rediscovered Stirling engines are external-combustion types: Heat is applied to a working fluid *outside* of the engine proper.

FOUR-CYCLE ENGINES

Internal-combustion piston engines are divided into two types; the 4-cycle engine is the most popular. All Briggs & Stratton products employ this type, as do the great majority of machines built by Clinton, Tecumseh, and Kohler.

The operating sequence in this engine is shown in Fig. 1-2. During the *intake stroke*, the piston moves down the bore. Since it fits quite tightly in the cylinder, it leaves a partial vacuum behind it. At the same time, the intake valve is opened by the camshaft. Air and fuel are drawn into the chamber in response to the vacuum, charging the cylinder.

The piston reaches the bottom extremity of travel (bottom dead center) and comes to a momentary stop, reverses direction, and travels upward in the *compression stroke*. Meanwhile, the intake valve has closed. The piston compresses the charge above it to one-seventh or one-eighth of its original volume. The effect makes the mixture more combustible by concentrating its energy into a smaller space.

Just before the piston reaches top dead center, a spark jumps across the plug gap and ignites the air and fuel, generating tremendous pressures in a very short time. Exhaust gases and superheated air drive the piston down in a *power stroke*.

As the piston approaches bottom dead center again, the exhaust valve opens. Moving upward on an *exhaust stroke*, the

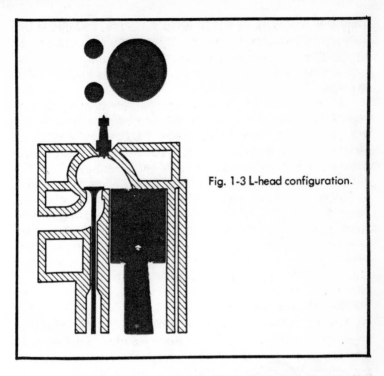

Fig. 1-3 L-head configuration.

piston drives the gases out of the cylinder and into the atmosphere. At this point, the sequence is ready to be repeated.

The name given to these engines—4-stroke-cycle—is understandable, considering their operation. Each up-and-down stroke of the piston corresponds to one of the four operating cycles. Since there are two piston strokes for each crankshaft revolution, the engine fires once every two turns of the crankshaft. In this country, we have shortened "4-stroke-cycle" to 4-cycle; in England, it's known as a 4-stroke.

The distinguishing feature of these engines is the valve mechanism. The valves are pushed ohn by the camshaft which is meshed with the crankshaft; there is one intake and one exhaust stroke per two crankshaft revolutions. In a like manner, the spark occurs every 720 degrees (twice 360) in large engines; but most lawnmower powerplants are driven at a 1:1 ratio and produce a phantom spark on the exhaust stroke.

The valves and cylinder bore form an **L** in the profile shown in Fig. 1-3. Reo and early Clinton engines had a valve on each side of the piston, giving them a **T** form. Because of the mechanical complexity (two cams were required), the **T**-head configuration has now become obsolete.

L- and **T**-head engines are known as side-valve or, more colloquially, as flat-head engines. Because the valves are alongside the piston, the combustion chamber is asymmetrical, with nooks and crannies which defeat complete burning of the fuel. These engines are also subject to *detonation*. (More on that later.) The alternative is the **I**-head, or overhead-valve (OHV) configuration. The valves stand directly above the piston as shown in Fig. 1-4. Briggs & Stratton, Continental, and other manufacturers have built such engines under military contract, but the Tecumseh (Lauson) type HH 150 is the first series built OHV engine for the civilian market. It is fitted to some of the better riding mowers and tractors.

Lubrication for these engines is provided by oil in the *crankcase*. Some designs are splash-lubricated; i.e., a scoop on the bottom of the rod splashes oil over the working parts (Fig. 1-5). Others employ a gear-driven oil slinger. Tecumseh

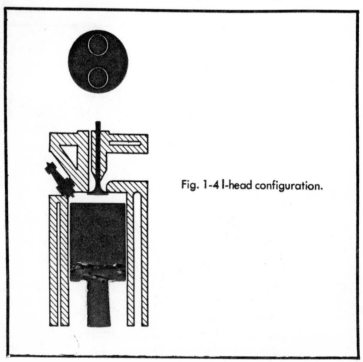

Fig. 1-4 I-head configuration.

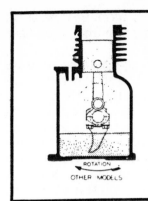

Fig. 1-5 Splash lubrication (Courtesy Clinton Engines Corp.)

ROTATION
OTHER MODELS

engines have always featured a piston-type pump—although this design is being supplanted with a geared pump. The lubrication circuit for the Kohler K662 twin-cylinder engines (Fig. 1-6), unlike most mower and tractor powerplants, is fully pressurized. There is no reliance upon splash to supplement pump output.

Fig. 1-6 Pressure lubrication   (Courtesy Kohler of Kohler)

TWO-CYCLE ENGINES

Two-cycle engines appear mechanically simple, but are sophisticated in concept. In the U.S., 2-cycle engines are built by Clinton, Chrysler, Jacobson, OMC (Lawnboy), and Techumseh (Power Products). Each of the 4 cycles of operation is present, but telescoped to give a power pulse for each crank revolution.

Figure 1-7 shows the working sequence. The crankcase is airtight, and forms part of the induction system. Events take place on both ends of the piston. Instead of valves, these engines breathe through ports in the crankcase and cylinder bore.

In the top drawing the piston is on the downstroke. It pressurizes the crankcase and forces air and fuel into the chamber through the open transfer port. The incoming charge, deflected by the dome on the piston crown, and pushes the exhaust gases out of the exhaust port. The intake and exhaust cycles overlap. In the middle drawing the piston moves toward top dead center, compressing the mixture in the chamber. At the same time the volume of the crankcase is increased, creating a slight vacuum. The leaf or *reed*, valve (Fig. 1-8) springs open—impelled by atmospheric pressure—letting fuel and air enter the crankcase. The bottom drawing shows the power stroke.

A controversy in 2-cycle design has long centered on the matter of *scavenging*. The bore has to be completely purged of exhaust residues during the intake cycle. The only practical way (in small engines at least) is to use the incoming charge as a ram to force the spent gases out. Of course, some of the fresh charge is lost; consequently, hydrocarbon emissions are high and fuel economy is poor.

The engine drawn in Fig. 1-7 employs a chamber that is, in technical terminology, *cross scavenged*. Economy is abysmal and the irregular shape of the piston crown gives rise to thermal stresses. The steep side of the deflector is cooled by the incoming charge while the shallow side suffers the brunt of combustion heat. The piston expands unevenly and, in extreme cases, can bind in the bore.

A small minority of lawnmower engines employ *loop scavenging*. Developed on the basis of the Schnurle patents in Germany, loop scavenged engines have flat (or slightly convex) pistons and two or more transfer ports (Fig. 1-9). The ports are angled to direct the charge upward where it

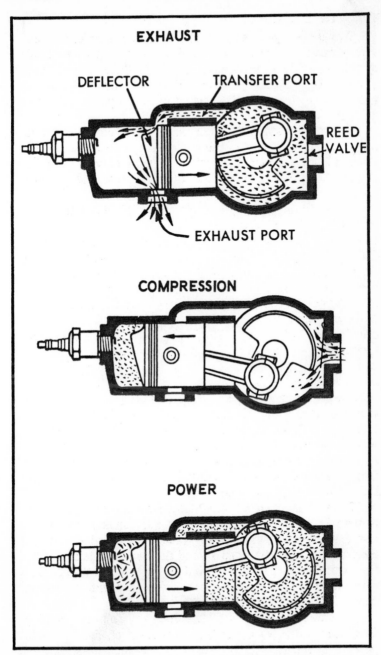

EXHAUST

DEFLECTOR

TRANSFER PORT

REED VALVE

EXHAUST PORT

COMPRESSION

POWER

Fig. 1-7 Two-stroke-cycle—cross-scavenged (Courtest Clinton Engines Corp.)

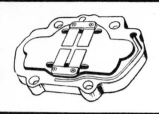

Fig. 1-8 Reed or leaf valve (Courtesy Clinton Engines Corp.)

strikes and rebounds from the roof of the chamber. The result is a miniature tornado whose edges displace the exhaust gases out of the bore.

Because the crankcase is part of the induction tract, 2-cycle lawnmower engines do not have an oil sump. Lubrication is afforded by mixing a small amount of oil with the gasoline. Oil drops out of the mix and coats the bearing surfaces. The usual oil/gasoline ratio is 16:1 or 24:1, although Jacobson engines operate happily at 32:1.

Total loss lubrication has several advantages, only two of which are simplicity (there are no pumps, slingers, or oil galleries), and cleanliness. Drawbacks are cost of operation, high hydrocarbon emission levels, and the inconvenience of having to mix oil and gasoline. This last point accounts for most of the customer resistance to 2-cycle engines.

POWER & TORQUE

The volume of an engine cylinder can be computed from bore and stroke measurements. The *bore* is the diameter of the

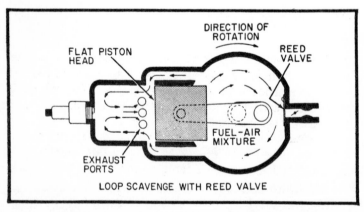

Fig. 1-9 Loop-scavenged engine (Courtesy Tecumseh Products Co.)

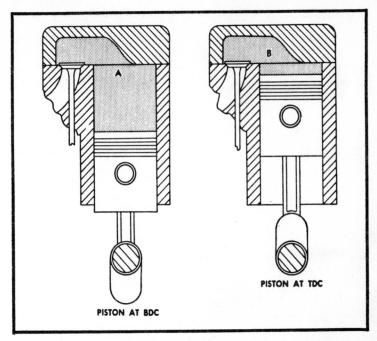

Fig. 1-10 Compression ratio is the ratio between A and B.

cylinder, and the *stroke* is the distance the piston travels from bottom, to top dead center—a fixed figure determined by the crankshaft offset or *throw*.

The swept volume, or *displacement*, is calculated by the formula

$$\frac{3.1416 \times \text{bore} \times \text{bore} \times \text{stroke}}{4} \times \text{number of cylinders}$$

The Lauson 4-hp series has a 2.5 in. bore and a 2.25 in. stroke. Using the formula, displacement is 11.04 cubic inches. To convert to metric units (something we will be doing more of in the future, multiply by 16.387. This particular engine has a displacement of 181 cc (cubic centimeters).

Displacement is a kind of touchstone for engine evaluation. When combined with other values—horsepower, fuel consumption, weight—it can give you a very good idea of the design philosophy.

Knowing the displacement of an engine in combination with compression ratio can give you an idea of power capability. All engines have a certain clearance volume above the piston (that area occupied by the spark plug) at top dead center. The ratio of the volume of the cylinder with the piston at bottom dead center (displacement plus clearance volume), to the clearance volume, is the *compression ratio*. The higher the ratio, the more the mixture is squeezed together, and the greater is the developed power. The pressure rise after ignition is three to four times the initial compression pressure. Unfortunately, compression ratios are sharply limited by the type of fuel available commercially and by combustion chamber shape. Lawnmower engines have compression ratios of between 7 and 8:1.

The heart of the engine is the combustion chamber, where gasoline and air are combined and ignited to produce an explosive reaction. Because it takes time for the pressures to reach their maximum values, ignition begins a few crankshaft degrees before top dead center. We say the ignition is *advanced*. A ball of blue flame expands outward from the spark plug tip at a rate of 20 to 300 feet per second. The rate of expansion depends upon the strength of the air/fuel mixture and the amount of turbulence present. Ideally, it should be as rapid as chemically possible. While we talk of this process as an explosion, it is more accurate to consider it a species of controlled oxidation or burning.

Near the end of the combustion process the unburned fuel and air at the extremities of the chamber are subject to increasing heat and pressure. Under certain conditions this residual mixture can spontaneously explode—a phenomenon known as *detonation*. Gas pressures rise 200–300 psi over normal. The rate of rise is so rapid that the connecting rod, cylinder, and cylinder head can go into sympathetic vibration. You may be able to hear detonation as a *ping* quite destinct from other engine sounds.

Detonation in lawnmowers can be controlled by setting the ignition advance to factory specifications and by using *regular*, instead of low-octane white gasoline. Keeping the chamber reasonably free of carbon accumulations should also help.

Pinging can be held to a minimum by mixing oil and fuel for 2-cycle engines in the correct proportion. Too much oil will worsen the detonation problem, more than using too little. Oil lowers the octane rating of the fuel and produces a hot flame.

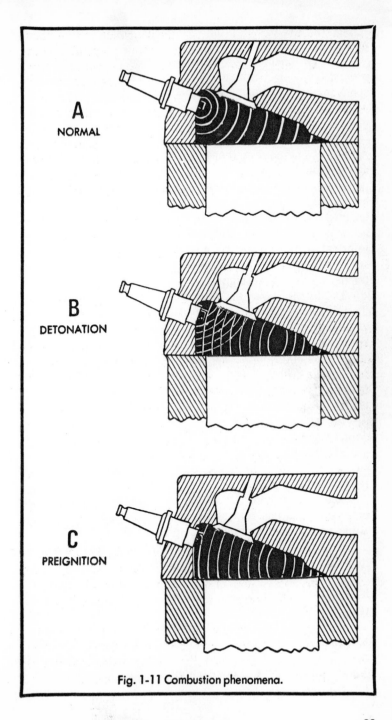

A
NORMAL

B
DETONATION

C
PREIGNITION

Fig. 1-11 Combustion phenomena.

Adjust the carburetor properly; a slightly rich mixture is preferable to keep the flame front moving.

Another phenomenon of abnormal engine operation is *preignition*, or ignition that occurs before normal, spark-induced ignition. The piston is caught between flywheel momentum and rising chamber pressures. Ultimately, the piston will seize in the bore, or melt.

The most common source of preignition is a local hot spot. A flake of carbon or a "hangnail" thread, detached from the spark plug boss, is usually the culprit. Two-cycle exhaust ports, especially those with bridges across them, can become hot enough to cause preignition. In general, lawnmower engines are so conservatively rated that preignition does not occur unless there is a failure in one of the support systems. Lubrication breakdown, cooling system failure (grass in the fan intake or clogged fins), or loss of governor control, raise temperatures to the preignition range.

The terms *torque* and *horsepower* relate to the output of the engine. Both have to do with the force exerted by expanding gases on the piston crown, and furnish a figure known as the *mean effective pressure* or—when calculated from horsepower—*brake mean effective pressure*.

Torque merely means "twisting force." When you open the lid of a jar you exert torque (Fig. 1-12). In the mechanical trades, torque is measured in *foot-pounds*, in terms of so many pounds of effort exerted on a bar one foot long. The torque demonstrated in Fig. 1-13 is 10 foot-pounds. Specifications for critical fasteners are given in terms of so many foot-pounds (ft-lb) of torque. In some cases the specification is in inch-pounds (inch-pounds = 12 × ft-lb).

The piston and rod assembly applies torque to the crankshaft journal. How much, and at what speed the torque value peaks, are functions of displacement and brake mean effective pressure. In practical terms, torque is a measure of the engine's ability to handle loads without bogging down.

While torque is an *instantaneous* measurement, horsepower is *energy expended over time*. One good horse is supposed to be able to lift 200 pounds 165 feet in 1 minute. Or saying it in another way, one horse develops the energy to raise 33,000 pounds (165 × 200) 1 foot in 1 minute. Whether or not this is true of horses is irrelevant; it has become the standard measurement of an engine's ability to do work (exert force through time).

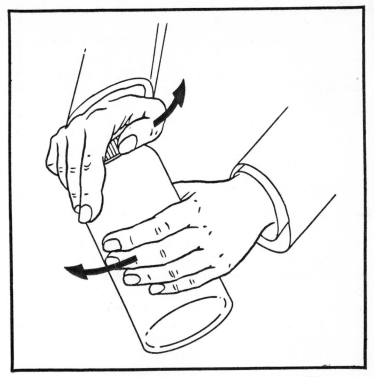

Fig. 1-12 Applying twisting effort, or torque, to a can lid.

Horsepower can be calculated with some precision, but the proof is to put the engine to an actual test. Figure 1-14 is a much-simplified drawing of a Prony brake. Engine power is measured on the scale in terms of pounds. Of course, this example of nineteenth-century engineering has long since been superseded by modern machines. Today, horsepower is

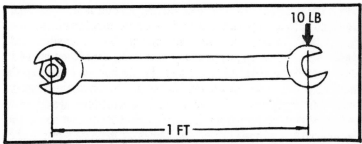

Fig. 1-13 Applying torque with a wrench.

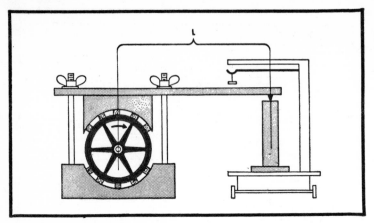

Fig. 1-14 Prony brake in simplified form.

measured electrically by resistance against a generator. But the Prony brake is useful because it gives us a formula;

$$bhp = \frac{2\,\pi\,lnw}{33,000}$$

bhp = brake horsepower
l = length of the arm in feet
n = crankshaft rpm
w = load in pounds on the scale
Suppose the length of the arm is 1 foot, rpm is 4000, and the load is 5 lb. Subtitution in the formula gives

$$\frac{2 \times 3.1416 \times 1 \times 4000 \times 5}{33,000} = 3.8 \text{ brake horsepower}$$

The Lauson engine in Fig. 1-15 develops a maximum of 6 ft-lb of torque at 3000 rpm. At an engine speed of 3600 rpm, the horsepower peaks. You may wonder why the horsepower (hp) continues to climb after the torque peak, since torque increases with *brake mean effective pressure* (bmep). As engine speed goes up, the pressure tends to drop because of a loss in "breathing" efficiency. The engine does not have time to develop a full charge. Horsepower, on the other hand, continues to climb because it is, in part, dependent upon engine speed. Eventually, the *hp* curve will falter as *bmep* values fall lower and friction increases. These factors can be compensated for by good design; the practical limit to

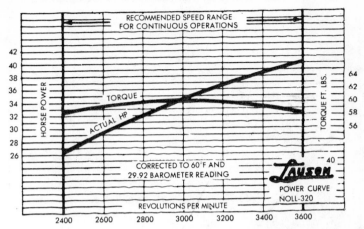

Fig. 1-15 Lauson power curve (Courtesy Tecumseh Products Co.)

horsepower is only determined by the strength of the connecting rod.

For comparison, look at the curves for the JLO-Rockwell engine (Fig. 1-16). The torque curve comes late in the speed band—past the rated speed of the Lauson engine—and is somewhat peaked. This engine operates most effeciently in a narrow speed band near the top of the curve. Horsepower rises dramatically between 3000 and 6500 rpm. The JLO works best when run hard. With slightly more than twice the displacement of the Lauson, it develops 35 hp.

The Lauson is an industrial powerplant intended for long and uneventful service. The flat torque curve means it will accept loads without faltering. Low specific output combined with a curve that is still climbing at the maximum governed speed means that the parts remain understressed. Designed for snowmobile service, the JLO's forte is longevity rather than flashy performance.

LUBRICATION

Motor oil reduces friction by interposing a liquid film between the bearing surfaces. As long as this film remains intact, there can be no wear, and frictional losses become a function of the resistance of the oil to shear. A plain bearing—e.g., a bronze bushing supporting a shaft—wears only during startup. Once the oil circulates, the rotation of the shaft tends to float it away from the bushing.

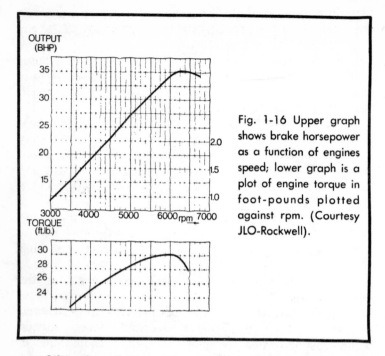

OUTPUT
(BHP)

TORQUE
(ft.lb.)

Fig. 1-16 Upper graph shows brake horsepower as a function of engines speed; lower graph is a plot of engine torque in foot-pounds plotted against rpm. (Courtesy JLO-Rockwell).

Oil is also a fairly good compression sealant between the piston rings and the bore. An engine which has been out of service for a long time will have a dry bore and hence very low compression. Another function is heat removal. On 4-cycle engines, a high percentage of the cooling is supplied by the oil in the sump. Low oil levels do not cause engine failure because of lack of lubrication; failure comes because the oil exceeds its critical temperature (usually about 275°F) and carbonizes. Instead of a lubricant, the oil becomes an abrasive which further increases temperatures.

In 4-cycle applications, oil should be able to resist combustion by-products. Each gallon of gasoline consumed produces a gallon of water, 0.1 gallon of unburned gasoline, and a broad spectrum of acids in the hydrogen, nitrogen, and sulfur families. In addition, lead salts, free lead (a gallon of "regular" contains about 1.0 cc of tetraethyl lead), resins, soot, carbon, and oxides of carbon and nitrogen are produced. A gasoline engine is a chemical plant with a surprisingly varied output.

The acids manufactured by this plant erode the bore and bearing surfaces. It is generally agreed that upper bore wear is primarily a result of the action of acid, although raw gasoline washing off the oil film also plays a role. Water can form a mixture with oil; in extreme cases this mixture becomes a reasonable facsimile of mayonnaise and causes rust. Gum and varnish residues combine to cause the rings to stick in their grooves. Carbon particles circulate about, ripping into bearing surfaces.

Modern oils have additives to combat most of these chemical menaces. Other additives increase the viscosity range or "pourability" at low temperatures, and others resist aeration. Detergents are additives which keep carbon and sludge particles in suspension; it is better for these particles to circulate than to fix themselves to bearing surfaces.

Additive mixes are specially formulated for 2-cycle engines. Resistance to combustion by-products is not as important as clean burning and good lubrication at high temperatures. Currently, the American Petroleum Institute is evolving a set of standards for air-cooled 2-cycle engines. Standards already in effect for liquid-cooled motors have enabled some outboards to use a gasoline-to-oil blend as lean as 50:1. Air-cooled engines undergo much higher temperatures, and these WC-T oils are inadequate for them—unless factory specifications allow it. You should always use the oils designated by the factory for any engine. API standards specify weight, viscosity range, and service profile. The highest rating in force is for automotives with 6000-mile change intervals. It is known as API-SE (service severe). It is interesting to note that Kohler does not recommend this oil because it lubricates so well that it prevents the rings from seating!

In general, it is best to avoid over-the-counter additives; there is no policing in the industry and some of these additives aree ree *positively harmful*. Those that are beneficial can be purchased more cheaply already mixed in quality oil.

COOLING

Normal combustion temperatures are double that of the melting point of aluminum. Friction—an engine may be considered one vast friction surface—also produces heat. These factors make it necessary to provide some form of cooling.

Although air is not a good cooling medium, since it has about 3500 times the resistance to heat transfer than cast iron, it has advantages for small engines. In a small air-cooled engine, the fan is integral with the flywheel and its output is roughly proportional to engine speed. Dropping the engine speed (and, therefore, fan output) after stressing the engine for a period of time will cause overheating. Air is taken from the flywheel hub area and discharged into sheet-metal shrouds which conduct it over the cylinder head. To make heat transfer more efficient, the heads are finned. The fins increase the area of metal exposed to the air blast and further tend to dissipate heat through their sharp edges. Two-cycle engines are usually more heavily finned than their 4-cycle counterparts. Thermal loads are higher (the 2-cycle fires every revolution) and there is no reservior of oil in the crankcase to help in cooling.

About all the owner can do to make sure his machine runs coolly is to keep the shrouds intact and periodically clean the fins as shown in Fig. 1-17.

EXHAUST EMISSIONS

Emissions from small engines have been the subject of considerable controversy. How should the emissions be

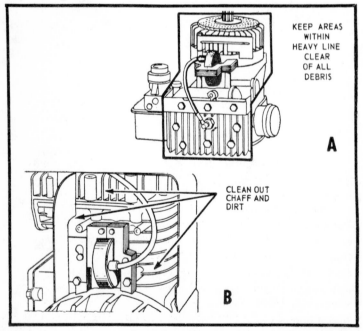

KEEP AREAS
WITHIN
HEAVY LINE
CLEAR
OF ALL
DEBRIS

A

CLEAN OUT
CHAFF AND
DIRT

B

Fig. 1-17 Areas to keep clean (Courtesy Briggs & Stratton)

measured? The operator of a lawnmower, rototiller, composter, edger, or the like would appear to be exposed to higher concentrations of pollutants than if he were driving a car. Some carbon monoxide is absorbed by microorganisms in the soil—but just how much is unknown. And it is difficult to estimate the use pattern of small engines, which is important both from the standpoint of pollutants released from individual engines and from the aggregate.

Certain broad bodies of information on the subject are forming. The SAE (Society of Automotive Engineers) *Small Engine Committee* has estimated the 1971 population of lawn and garden engines in this country to be 38,700,000. This figure is based on 1968 production records and an estimated 5-year lifespan. Less than 10% of these were 2-cycles. Assuming an average annual use figure of 50 hours, we can, by measuring engine emissions, get a fairly good idea of the relative pollutant output of lawnmower engines.

There are three types of exhaust gas emissions of central concern at present. These are HC (hydrocarbons consisting of unburned gasoline and oil), carbon monoxide (CO and a broad collection of oxides of nitrogen lumped together as NO_x. The SAE study produced these figures:

4-CYCLE LAWN EQUIPMENT	TOTAL EMISSIONS (TONS/YEAR)
HC	37,800
CO	664,000
NO_y	10,400
2-CYCLE LAWN EQUIPMENT	
HC	24,200
CO	57,000
NO_x	161

Remembering that 2-cycles account for only 6.9% of the total number, the emission figures are surprising. These same engines produce 64% of the total HC emissions. This can be traced to poor scavenging and the presense of oil in the fuel. At the same time, oxides of nitrogen are very low—only 1.5% of the total. Oxides of nitrogen are the most difficult pollutants to control; and at present, some of the exhaust gases are recirculated to quench the combustion temperatures. Obviously, 2-cycles do this automatically. Co figures lie somewhere between those for both engine types. The 2% or so bias against 2-cycles can be easily explained as a measurement error.

Lawn and garden engines contribute little to the total air pollution picture. They release only 0.2% of the total HC, 0.7% of the CO, and less than a 0.1% of the NOx.

The most intensive study of individual small engine emissions has been made under government contract at the Southwest Research Center (San Antonio, Texas) under the guidance of Charles T. Hare and Karl J. Springer. The engines used were the Tecumseh (Power Products) AH520, Briggs & Stratton 92908 (vertical-shaft rotary lawnmower), 100202 (horizontal-shaft reel mower and edger), Kohler K842 (gas-powered lawnmower) and Wisconsin S-12D (garden tractors). The emission data was gathered during numerous runs at several throttle positions and loads. Carburetors were adjusted to best lean power. The following data is presented on the basis of grams per horsepower per hour.

HYDROCARBONS	ENGINE APPLICATION/TYPE	BRAKE SPECIFIC EMISSIONS. g/hp hr
(EXHAUST ONLY)	LAWN & GARDEN/4-STROKE	23.20
	LAWN & GARDEN/2-STROKE	214.00
	MISCELLANEOUS/4-STROKE	15.20
CO	LAWN & GARDEN/4-STROKE	279.00
	LAWN & GARDEN/2-STROKE	486.00
	MISCELLANEOUS/4-STROKE	250.00
NOx AS NO1	LAWN & GARDEN/4-STROKE	3.17
	LAWN & GARDEN/2-STROKE	1.58
	MISCELLANEOUS/4-STROKE	4.97
RCHO AS HCHO	LAWN & GARDEN/4-STROKE	0.49
	LAWN & GARDEN/2-STROKE	2.04
	MISCELLANEOUS/4-STROKE	0.47
PARTICULATE	LAWN & GARDEN/4-STROKE	0.44
	LAWN & GARDEN/2-STROKE	7.10
	MISCELLANEOUS/4-STROKE	0.44
*SOx AS SO2	LAWN & GARDEN/4-STROKE	0.37
	LAWN & GARDEN/2-STROKE	0.54
	MISCELLANEOUS/4-STROKE	0.39

*Calculated on basis of 0.043% fuel sulfur content by weight. (Courtesy Southwest Research Institute Gov. Contract No. EHS-70-108).

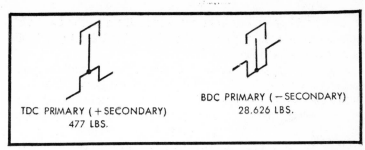

TDC PRIMARY (+ SECONDARY)
477 LBS.

BDC PRIMARY (− SECONDARY)
28.626 LBS.

Fig. 1-18 Primary and secondary forces.

VIBRATION

One of the facts of lawnmower life is vibration. The intrinsic vibration of a single-cylinder engine is often amplified by the deck and blade to galling proportions. Indeed, an out-of-balance blade can literally shake an engine to pieces.

The reciprocating parts (piston, wristpin, and the upper half of the connecting rod) have inertia—they resist a change in velocity or direction. If the rod were to come apart, the piston would continue to travel until it collided with the roof of the combustion chamber or the crankcase. Inertial vibration is called *primary shaking force*. It occurs at TDC and BDC (on the axis of the bore) twice during each revolution when the piston is restrained to a dead halt.

The connecting rod also generates vibration. These *secondary shaking forces* occur at TDC where they add to the primary force and at BDC where they subtract from the primary force. The secondary forces modify the primary force by a margin of 25% in both extremes of piston position. The formula for the primary shaking force is:

$$F = 142 \times 10^{-7} WN^2 S$$

where
F = primary shaking force
W = weight of the reciprocation parts
N = rpm
S = stroke

Let us assume that the reciprocating weight is 0.74 lb, the stroke is 2.25 in. and the rpm is 4000. These figures approximate the values for small mower engines. By formula, the primary shaking force is 383.4 lb. Subtracting 25%, or 95.9 lb, from the BDC and adding it to the TDC complement gives us 479.3 lb upward and 287.5 lb downward. These forces are portrayed in Fig. 1-18.

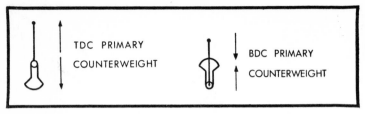

Fig. 1-19 The effect of a crankshaft counterweight.

If the TDC force persisted—say, for 10 seconds—the lawnmower (weighing an estimated 70 lb) would accelerate to a velocity of 21,760 ft/sec or almost 15,000 miles an hour. Fortunately, the impulse is only momentary and is partially canceled by the BDC complement. But you can appreciate the importance of keeping the mounting bolts secure.

Some of the vibration is attenuated by the crankshaft counterweight. The counterweight compensates for the weight of the crankpin and part of the reciprocating mass. At both ends of the stroke, the counterweight sets up forces in opposition to the shaking forces (Fig. 1-19). However, balance cannot be perfect. If the counterweight exactly matched TDC forces, it would generate a 25% or so imbalance at BDC. And at midstroke the counterweight adds a transverse component with nothing to oppose—the engine would move sideways. As a compromise, most engines are counterweighted to 50−60% of the reciprocating mass.

An attractive solution is to add a second cylinder in 180° opposition to the first. The Power Products crankshaft shown in Fig. 1-20 does not require counterweights—the crankpins and reciprocating masses are balanced. Both forces (TDC and BDC) are canceled—both cylinders reach the ends of the stroke together. But, all is not perfect; the crank tends to skew against the main bearings at the limits of piston travel. This phenomenon is known as a *rocking couple* and becomes more serious as the cylinders are further offset.

The crankshaft shown was used in railway maintenance vehicles. Unlike outboard and snowmobile manufacturers, lawnmower makers have tended to remain loyal to the single-cylinder concept.

The Kohler twin-cylinder models have made small inroads into the market as have singles with vibration-compensating gears or dummy connecting rods. Briggs & Stratton has

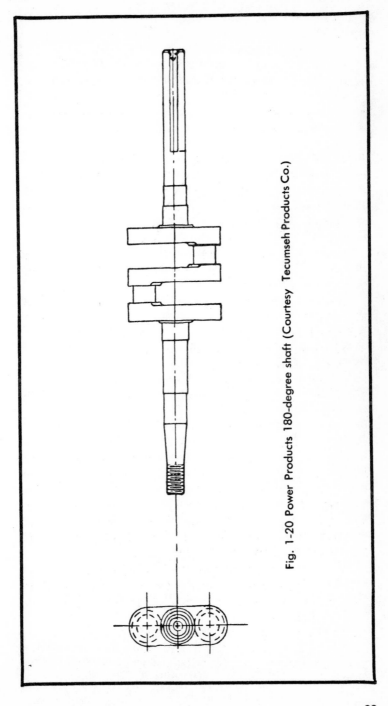

Fig. 1-20 Power Products 180-degree shaft (Courtesy Tecumseh Products Co.)

employed both methods, while Kohler and Tecumseh employ contrarotating counterweights. How these weights compensate for primary forces is shown in Fig. 1-21.

In the long run, the vibration problem may be solved by the Wankel rotary engine, which has no reciprocating masses. Electric power might also provide a solution. We are in the second generation of self-contained electric mowers, and their performance is competitive with gasoline plants.

Engine Construction

The piston is made of cast aluminum and is usually ground to an oval shape. The largest diameter is at the wristpin. As the piston expands under combustion and frictional heat, the oval fills out into a true circle for full contact with the bore. Without the benefit of *cam* grinding, the piston would "slap" when cold. Some are ground with a slight taper at the crown, because the crown runs hotter and expands more than the body of the cylinder (see Fig. 1-22).

Several manufacturers offset the wristpin slightly. This offset cants the piston at top and bottom dead centers and reduces the noise level. From a mechanic's point of view, the offset means that the piston must be installed in accordance with the factory match marks (Fig. 1-23).

The piston must be as light as possible consistent with reasonable durability. What constitutes "reasonable durability" is subject to dispute, but when pressed to the wall, manufacturers quote figures of 500—750 hours between overhauls. The crown is relatively thick to withstand the thermal stresses, and the wristpin journals are generous. But the *skirt* (the area below the pin) is quite thin. Many pistons have skeletal skirts for weight and friction reduction.

Wristpins used in lawnmower engines are known as the *full-floating* type. They oscillate on the small end of the rod and—once operating temperature is reached—on the piston bosses. In a few instances, the pin is supported on a bronze bushing at the connecting rod. Rarer yet is the use of needle bearings: the wristpin is secured by a pair of spring clips which should be renewed with each overhaul. Should one of these clips (Fig. 1-24) fail, the pin will move into contact with the cylinder bore. You can imagine the results.

The wristpin usually outlasts every friction surface on the engine. Elongation of the rod at the small end, or pounding damage to the bushing, are more common than pin failure.

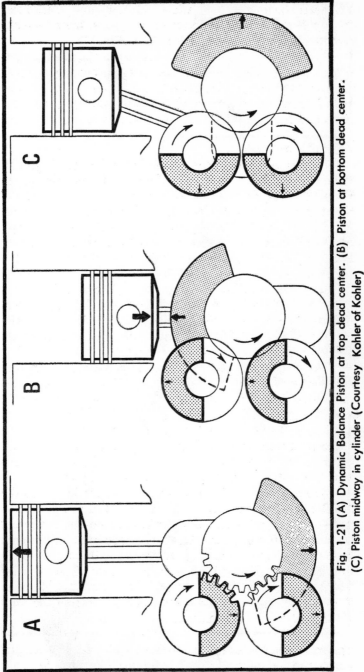

Fig. 1-21 (A) Dynamic Balance Piston at top dead center. (B) Piston at bottom dead center. (C) Piston midway in cylinder (Courtesy Kohler of Kohler)

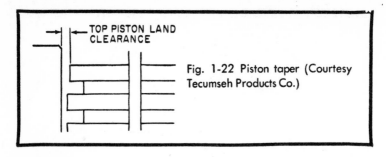

Fig. 1-22 Piston taper (Courtesy Tecumseh Products Co.)

TOP PISTON LAND CLEARANCE

The rings seal, cool, and, in 4-cycles, lubricate the piston. Surprisingly enough, a good third of the piston's heat load passes through the rings to the cylinder. Sealing depends on initial ring tension or "spring," and on controlled gas pressure vented behind the ring that forces it more tightly against the cylinder wall. The compression ring in Fig. 1-25 has a 45° bevel on its upper and inner circumference to encourage dilation. Other compression rings, especially in 2-cycle service, may have a rectangular cross section.

Two-cycle engines are fitted with two compression rings. Four-cycles require an oil ring and may have a scraper. The scraper (Fig. 1-25) prevents excess oil from entering the chamber. Many small engines still employ a scraper ring, although it must be considered obsolete. A second compression ring can give the same oil control and does a better job of

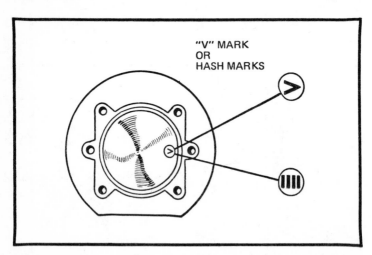

"V" MARK OR HASH MARKS

Fig. 1-23 Alignment marks for piston offset (Courtesy Tecumseh Products. Co.)

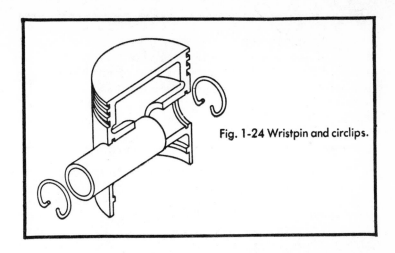

Fig. 1-24 Wristpin and circlips.

sealing. The oil ring is generally a one-piece affair, with bridges to distribute oil along the bore. The lower part of the cylinder is inundated by splash from the crankcase. Excess oil is channeled through holes in the ring to the back of the groove where it is drained through vents. Unlike other rings, those for oil do not have a symmetrical shape.

Four-cycle rings are free to rotate in their grooves. One study has shown that they revolve at about 80 rpm. Rotation is desirable since it helps prevent varnish and carbon buildup. Two-cycle engines often have pegged rings as shown in Fig. 1-26. If the rings were free to move, the ends could snag on the ports.

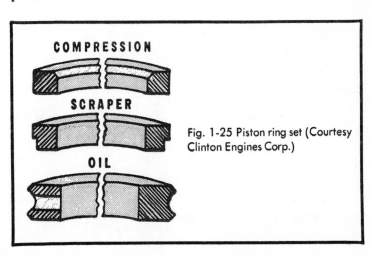

COMPRESSION

SCRAPER

Fig. 1-25 Piston ring set (Courtesy Clinton Engines Corp.)

OIL

Rings are made of fine-grain cast iron. This material has good heat conductivity and is compatible with cast iron bores; an unusual situation—most metals do not take kindly to a bearing made of the same material. The compression rings may have a chromed face to improve wear characteristics and reduce acid damage. However chromed rings must not be used in chrome bores.

The connecting rod may be made of forged steel (rare in lawnmower service), bronze (Jacobson and early Power Products), or aluminum. The latter has the advantage that it can be used as a bearing surface at both ends of the rod. Briggs & Stratton pioneered this concept and it has been followed by other small-engine builders—and by the makers of Jaguar cars. Aluminum fatigues easily and these rods are correspondingly larger and heavier than one would suppose, for the stresses involved.

The rod is split at the big end for installation over the crankpin. The alternative, built-up crankshaft (bolted or pressed together), remains the prerogative of motorcycle manufacturers—it is entirely too expensive for small industrial engines. If possible, the rod is split on the horizontal plane. Very large crankpins combined with narrow bores dictate that the rod be parted on an angle (Fig. 1-27); otherwise, it could not be withdrawn from the bore. Stresses are nonsymmetrical in this design and bear heavily on the parting line.

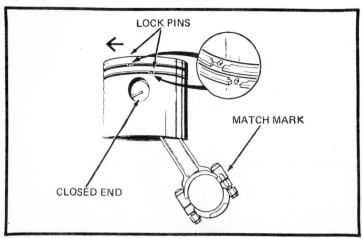

Fig. 1-26 Two-cycle rings with lock pins (Courtesy Tecumseh Products Co.)

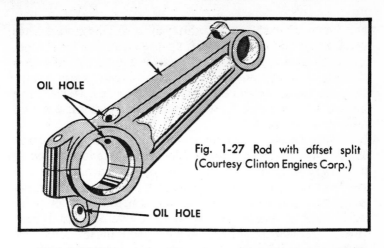

OIL HOLE

OIL HOLE

Fig. 1-27 Rod with offset split (Courtesy Clinton Engines Corp.)

The bearing may be the rod material itself, a precision insert, or a row of needles. Good-quality 2-cycles feature forged steel rods with needles (Fig. 1-28). Wisconsin engines found on commercial mowers employ replaceable inserts. In any event, the big-end eye must be precision-reamed. The individual rods are assembled at the factory for finishing. This means no two caps are interchangeable and the cap must be installed in its original position relative to the shank. The big

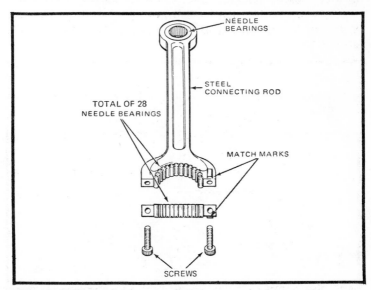

NEEDLE BEARINGS

STEEL CONNECTING ROD

TOTAL OF 28 NEEDLE BEARINGS

MATCH MARKS

SCREWS

Fig. 1-28 Power Products needle-bearing rod (Courtesy Tecumseh Products Co.)

end has embossed marks which must be aligned during assembly.

Most manufacturers secure the rod caps by hex-head screws which are threaded into the shank; some Power Products models thread the caps. Because rod material is by no means the strongest for screw threads, the Lauson practice of using a separate bolt and nut appears superior. On the other hand, lawnmower engines rarely "throw" rods in normal service. Most failures are caused by inadequate lubrication or a runaway governor.

The rod fasteners are subject to elongation and compression. To prevent the nut from loosening, a lock plate (Fig. 1-29) or Ny-Loc fasteners are used. The latter is a patented system consisting of nylon buttons cast into the fastener body. When tightened, the nylon deforms to match the thread profile. These fasteners should be renewed with each overhaul.

The crankshaft is the most critical single part of the engine and the most expensive. Made of cast iron, the bearing surfaces are nitrided or induction-tempered for long wear. Straightness for lawnmower service is held to ±.001 in. The diameter of the big end journal is held to 0.00015 in. with needle bearings; twice that is allowed for plain bushings. The main journals can be somewhat less precise because the stresses are more evenly distributed. In general, lawnmower engines employ plain bushing main journals with a quality gradient derived from ball bearings on the power takeoff end, to full ball or taper bearings on the other end. Blade-induced thrust is normally taken by the upper, or magneto, end of the main journal.

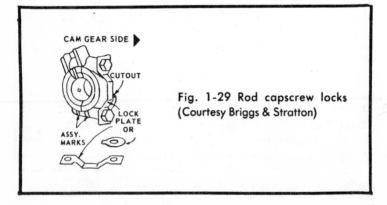

CAM GEAR SIDE ▶

CUTOUT

LOCK PLATE OR

ASSY. MARKS

Fig. 1-29 Rod capscrew locks (Courtesy Briggs & Stratton)

The lay of the crankshaft determines engine classification. Vertical-crankshaft engines are widely used with rotary lawnmowers, while the horizontal configuration is standard for edgers and reel mowers. The vertical-shaft engines are steadily becoming more popular in riding mowers since power transfer to the blade is more direct.

Most rotary and riding mowers feature 1:1 blade coupling through an adapter which is keyed to the crankshaft stub. The adapter is almost always intended to slip under sudden impacts, on the theory that the crankshaft and blade will be protected from damage. However, a solid collision will invariably bend the blade and crankshaft. For this reason, commercial mowers typically drive the blade by means of a belt; the engine can turn slower for a given blade tip speed and impact damage is restricted to the blade and blade shaft.

The flywheel is a storehouse for kinetic energy. It collects energy on the power stroke and releases it during the other cycles. An engine with a heavy flywheel will be easy to start and will munch through heavy grass without bogging down. The better engines employ cast iron or massive aluminum wheels. Light wheels reduce the weight (and cost) of the package and allow quick acceleration. A clever design engineer can balance a light wheel with a responsive governor and get more or less satisfactory operation. The cost—engineering is always a matter of tradeoff—is a nonuniform rpm curve.

Most proprietary vertical crank engines—that is, those which are wholesaled to an independent manufacturer for boltup into a lawnmower—have tiny flywheels. The blade mass supplies inertia for starting. This strategem works since a lawnmower should not be operated with a loose or missing blade. But on certain belt-driven machines with big flywheels it does lead to problems. Yazoo engines in particular have entirely adequate flywheels and will run without the belt; but some of the inferior imitations do not have enough mass clustered around the crankshaft. The blade should provide it, but in practice inertia is lost as the belt slips during starting. Consequently, these engines spit back—they bite the hand that pulls the cord—and are frequently in the shop. There is little that the mechanic can do short of installing a higher-mass flywheel.

The flywheel must be secure; a slight misalignment between it and the shaft will upset the ignition timing. And an undertorqued nut can lead to flywheel breakage in the event of sudden impact.

The camshaft works the valves through a pair of eccentrics lifting against tappets (cam followers). It turns at half crankshaft speed and must be initially timed to open the valves in sync with the piston. All contemporary engines have reference or timing marks on both parts, but you may have to look carefully to find them. The crankshaft gear mark may be in the form of a chamfered tooth, keyway (Tecumseh), or a punch mark. Camshaft gears are punched or indented on either side of the meshing tooth as shown in Fig. 1-30. If you cannot find the marks, make your own before final disassembly.

Because of its reduced speed, the camshaft is a traditional power takeoff source. Wisconsin and other large engines use it to drive the magneto. It drives the oil pump or slinger and most manufacturers offer a propulsion option (Fig. 1-31); the latter makes possible positive wheel drive without the mower manufacturer having to do much engineering. Most mower factories are assembly plants (rather than design & development firms).

Valves take a beating from heat and corrosion. The exhaust valve suffers more than the intake valve because it does not have the benefit of charge-cooling. Valve metals vary, but the better engines use stainless steel—at least on the exhaust side. A few high-quality engines have sodium-cooled

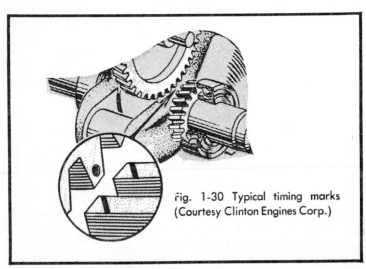

Fig. 1-30 Typical timing marks (Courtesy Clinton Engines Corp.)

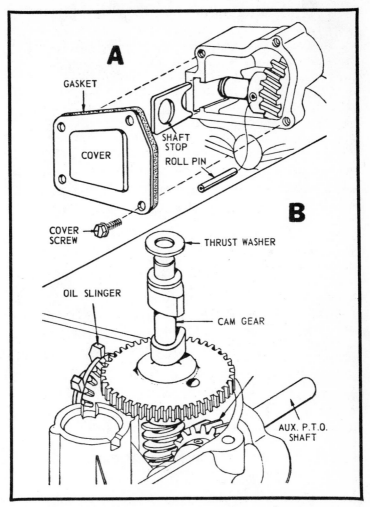

Fig. 1-31 Camshaft power takeoff (Courtesy Briggs & Stratton)

exhaust valves. The stem is hollow and filled with grains of this silver-white metal. At operating temperatures, the sodium becomes liquid and transfers heat from the valve head to the stem and valve guides.

Valve life can be increased by employing a valve rotator (Fig. 1-32). Normally fitted to the exhaust valve in long-distance engines, the rotator gives the valve a twist with each lift. The wiping action tends to keep the seat and valve face clean.

The majority of small engines have a valve face angle of 45°. Except for Briggs & Stratton intake valves, you will have to go back many years to find 30° faces, which were used in the mistaken belief that the shallow angle increased gas flow. Some operating clearance between the valve and the cam follower must be allowed to compensate for heat expansion. Broadly, the clearance is 0.010 in. for cast iron blocks and 0.012 in. or more for the aluminum variety. The difference is in the expansion rates of the two metals.

The valve springs pull the valves down as the cam turns past the high point of travel. Auto and motorcycle engines have a pair of concentric springs on each valve to eliminate high-speed shock waves. But lawnmower engines are understressed and get by with a single spring. In some examples, the exhaust spring has greater tension than the intake side—either from the use of heavier-gage wire or from the composition of the material.

Valve guides support the valves, keep them over the seats, and transfer their heat out to the cylinder fins. They are made of sintered bronze or cast iron. The guide is pressed into place with an interference fit, into the block. Very cheap engines sometimes run the valves directly on aluminum—the principle of the wonderful one-horse shay.

Cast-iron engines generally do not employ separate valve seats as such. Iron presents an adequate wearing surface for normal use. Light-metal blocks require ferrous seats to withstand the battering of valves slamming shut. Lawnmower engines have *inserted*—as opposed to cast-in—seats. Besides simplifying foundry work, inserted seats can be replaced in the field without a milling machine. On the other hand,

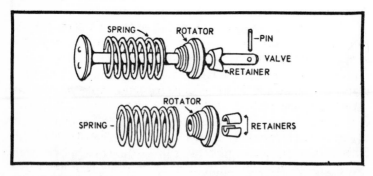

Fig. 1-32 Valve rotator for severe service (Courtesy Briggs & Stratton)

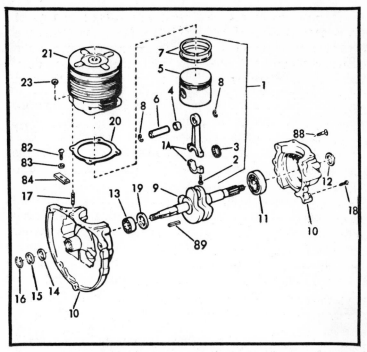

Fig. 1-33 Power Products engine in exploded view. The cylinder barrel is 21 (Courtesy Tecumseh Products Co.)

because inserted seats are held to the block only by the forces generated at the fit, they can pound loose. The problem becomes more severe as the engine is run at full load and the block metal reaches maximum expansion. The seat OD (outside diameter) is normally about 0.040 in. larger than the recess in the block. Replacement seats are secured by peening the block metal.

Of course, 2-cycle engines do not have the mechanical complexity (some would say wackiness) of cams, valves, springs, rotators, etc.

The term *block* is a general one, and includes both the crankcase and the cylinder. Both components may be cast in one unit, as illustrated in most of the drawings in this chapter. Power Products, OMC, and Jacobson 2-cycles employ a separate cylinder casting bolted to the crankcase (Fig. 1-33). Both of these castings are collectively termed the *block*. We speak of the crankcase and the cylinder barrel (or jug) individually.

The block may be of aluminum or cast iron. Cast iron is cheap, fairly easy to machine, and has dimensional integrity—that is, the bore stays round and the crankshaft and camshaft bosses remain parallel and true. On the debit side, iron is heavy enough to have been the favored material for anchors and sash weights. This is, of course, a disadvantage on any vehicle, whether the owner or the engine supplies the propulsion; and iron is a poor medium for heat transfer. Local hot spots tend to develop, and the valves are not cooled as they should be. Consequently, iron engines have low specific outputs.

Aluminum is light, has about four times the thermal conductivity of iron, and (properly alloyed) can double as a bearing surface. The material responds to precision foundry techniques, which means a minimum of machining and a maximum of thin, closely spaced fins. Compare the fins on a Power Products 2-cycle with the thick, almost globular fins on a Kohler or Wisconsin. Unfortunately, aluminum is fairly expensive and getting more so with the increase in energy and bauxite costs. In lawnmower and garden equipment service, aluminum engines just don't have the durability of cast iron. These engines are stressed more, partly because the superior cooling and design philosophy allows it; but the blocks tend to warp. How much of this warpage is due to poor design is a matter of conjecture. An aluminum block will not warp if provided with the proper stiffeners. Many utility engines appear to be cast from the same molds used for iron. The weight saving is dramatic; but durability is another issue.

While aluminum can support the main bearings and camshaft without the intermediary of a bushing or an antifriction bearing, it cannot withstand piston ring friction. An exception is Chevy's Vega engine which is cast in a high silicon content alloy. The bore is acid-etched so that the rings slide on an irregular surface whose peaks are composed of silicon crystals.

Most small engines employ an iron sleeve in the bore. The sleeve is first dipped in pure aluminum and then the block is cast around it. The iron and aluminum become molecularly bonded into a single unit. Heat transfer across the interface is good, although some bore distortion may be expected. These engines can be rebored in the field and fitted with larger pistons and rings.

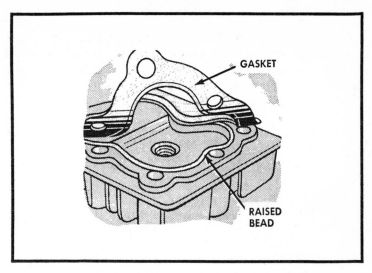

Fig. 1-34 Head gasket (Courtesy Clinton Engines Corp.)

Briggs & Stratton production is divided between cast iron, cast iron liner, and chrome-block engines. The latter technique is unique to Briggs. It consists of applying a chrome finish directly to the aluminum cylinder walls. The chrome is "back-etched" to provide a rough surface for oil adhesion; the cylinder is used with a tin-plated piston and plain iron rings. Bore life is excellent since chrome provides a hard-wearing surface and is almost immune to acid attack.

Four-cycle vertical-shaft engines have an oil sump or flange below the block. The flange supports the lower main bearing, camshaft, and oil slinger or pump. It carries mounting bosses to fasten the engine to the lawnmower deck. The industry has attempted to standardize mounting fixtures. Nearly all small vertical-crankshaft flanges have three small mounting holes, or slots, on an 8-inch circle. A 6⅞-inch hole is required to clear the sump depression and to provide access to the drain plug.

Although many 2-cycle engines have a cylinder barrel and head cast as a single entity, the usual procedure is to separate the two. The designer has no choice with side-valve 4-cycles; and a detachable head on a 2-cycle can make port profile and alignment somewhat less formidable.

The head is secured by four or more bolts which may be clustered in the exhaust valve area. The interface is sealed by a composition gasket (Fig. 1-34), although the gasket could be dispensed with between two closely milled aluminum surfaces. Blown head gaskets can usually be traced to failure by the owner to retorque the head after break-in, overtorquing, or in a few cases, poor design. There must be equal stress on the head fasteners when the engine is hot and the block has expanded.

Chapter 2

Tools

Lawnmowers are relatively austere machines with few of the engineering complexities we have come to associate with autos and home appliances. Ordinary hand tools suffice for most repair jobs. Special factory tools make the work go easier, but a little ingenuity can usually get you by.

The tool situation is clouded by metrification. Foreign mechanics must purchase U.S.-standard tools to service American products; foreign machine-tool orders are now specified in metric modules. That is, foreign tools are dimensioned on the metric scale directly, as if the inch-pound standard did not exist."

In the U.S., the Metric Conversion Act of 1972 died in the House Committee on Science and Astronautics, but recent indications are that it soon will be resurrected. Industry has responded by quietly shifting to the metric standard. GM's rotary engine was to have been metric from the drawing board up. The Ford plant at Lima, Ohio, is punching out metric-standard engines for the Mustang II. This is the first all-metric engine designed and built in the U.S. Caterpillar Tractor has also gone metric—but not in fasteners. The problem here is competing standards within the metric camp. It is further complicated by the generally looser tolerances specified by the Europeans and by the lower tensile strength of many metric fasteners.

But metrification is coming, either in the back door, by virtue of the efforts of export-conscious manufacturers, or openly by legislative action.

This means that the part-time or professional mechanic must purchase tools with an eye to their long-term utility. Wrenches, taps and dies, reamers, and precision measuring instruments may have to be duplicated when metric small engines arrive. Metric fasteners will have to be segregated from inch-standard sizes, since there is a real possibility for confusion. Codes stamped on the heads of fasteners to indicate their tensile strength will also have to be changed.

WHAT QUALITY TOOLS?

Some people collect tools as if they were jewelry or rare coins. For them, the value of the tool lies in the craftsmanship which went into it. Unfortunately, few of us can afford this kind of esthetism. The Thor minidrill is perhaps the most sophisticated ¼-inch drill motor available. It is half the size of a conventional motor, has a stator guaranteed for life, and employs full-wave rectification and a permanent-magnet. But all it can do is bore a ¼-inch hole in steel. A $7.98 cheapie may sputter and howl, but it will bore a ¼-inch hole in steel too.

The very best hand tools (Fig. 2-1) are made by firms like Snap-On, Blackhawk, New Britain, and Armstrong. These tools have a finish and heft which makes them a pleasure to use. The quality is such that you will wear out *before they do*. But the average mechanic might do better, at least in terms of the range of his tool inventory, to purchase less expensive brands.

Inexpensive, in this case, does not mean cheap imports sold in five-and-dimes or drugstores; most of their offerings are made of castings or welded-up stampings. In many cases, the steel is softer than the fasteners they are supposed to move.

The best choice, with respect to a use/cost factor, are tools made from molybdenum and vanadium alloys, forged for strength, and triple-chromed for ease of cleaning and rust protection. Look for known brands such as S-K, Husky, Wright, and Proto. Auto parts jobbers carry tools in this quality level, but you may be able to best their prices since these outlets do not discount tools. Major department stores such as Sears and Wards stock a large inventory of hand tools and are favorites of many mechanics.

Wrenches

Any tool collection should begin with a set of *open-end* wrenches. You will need sizes ranging from ¼ in. to 1 in., in increments of sixteenths of an inch. Some power lawnmowers employ $^{11}/_{32}$ in. and $^{19}/_{32}$ in. fasteners. Purchase wrenches for these as you need them.

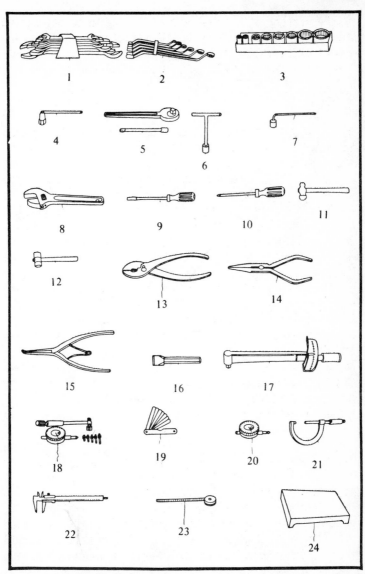

Fig 2-1A Standard tools for small engine work: 1—open-end wrench set; 2—box-end wrench set; 3—socket wrench set; 4—plug wrench; 5—ratchet handle and extension; 6—T wrench; 7—L wrench; 8—adjustable open-end wrench; 9—slot-head screwdriver; 10—cross-head screwdriver; 11—steel hammer; 12—plastic hammer; 13—pliers; 14—longnose pliers; 15—Circlip expander; 16—chisel; 17—torque wrench; 18—cylinder gage; 19—thickness gage; 20—dial gage; 21—micrometer; 22—vernier calipers; 23—tape measure; 24—surface plate.

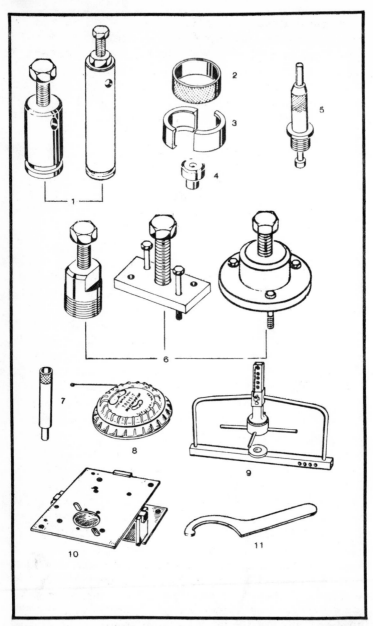

Fig 2-1B Factory tools for specialized jobs: 1—universal assembly jig; 2—retaining ring; 3—half-shells; 4—pressure piece; 5—timing gage; 6—flywheel extractor; 7—guide pin; 8—rpm indicator; 9—recoil-starter clamping jig; 10—universal assembly jig; 11—spanner wrench.

Box-end wrenches have closed jaws and give a superior grip. Shop teachers sometimes suggest that box-ends be used for initial loosening and final tightening. Working mechanics use box- and open-end wrenches pretty well interchangeably. The major advantage of a box-end is that it will hold a battered fastener which would ordinarily defeat an open-end.

Combination wrenches combine both jaw styles on the same shank. Because the sizes are the same, you have to buy twice as many wrenches to get a working spread. Combos are a luxury, but a very seductive one. Heft and balance is superior to either of the others and you have an always available alternate jaw. A beginner might do well to invest in a set of combination wrenches and forget about the others until needed.

The standard machine fastener has a hex (six-sided) head. *Socket* wrenches are available with six or twelve points, as are small box-ends. The advantage of the 12-point jaw is quick attachment; however, 6-point wrenches are stronger, more resistant to wear, and have more torque capability—you can apply greater pressure without deforming the bolt faces. In general, you should buy 6-point thin-wall sockets, and specify 12-pointers in the standard wall thickness. Socket reach comes in three sizes—short, medium, and deep-well. The medium is most versatile, and will handle 90% of lawnmower fasteners. Sparkplug sockets feature external flats; wrenches can be applied directly to the socket, and they usually have rubber inserts to grip the plug insulators. A $^{13}/_{16}$ in. deep-ell is entirely adequate and slightly cheaper for standard plugs. Miniplugs require a ⅝ or ¾ in. socket. Of course, sockets are of little use without a driver.

A *ratchet handle* (sometimes called a ratchet wrench) is an indispensable tool. Purchase a good name brand and take good care of it. All ratchet handles wear and eventually slip. Most established brands furnish repair kits consisting of faceplates, wheel, springs, and ratchets. The kits cost about half as much as the complete wrench. Sears & Roebuck is unique in that they replace worn ratchet handles free-of-charge. This policy is responsible for much of the goodwill mechanics feel toward the Craftsman line of tools. Other firms replace their wrenches if failure can be traced to faulty manufacture. In effect, the customer has to convince the retailer that he didn't use a cheater bar or a sledgehammer on the wrench. Then the customer has to wait

53

a month or so for the ompany to make a final decision on his case.

Ratchet handles and other socket drivers are grouped according to the dimension of the drive stud. Small engine work goes best with a ⅜ in. drive, although there are occasions when the heftier ½ in. drive is more appropriate. If you use your tools to work on your car as well, you might invest in a Husky Model CS 43. Although this is a ½ in. drive handle, it is as compact as most ⅜ in. models.

The size of the ratchet-wheel teeth determines the bite of the tool and the overall dimensions of the head. The latter is not critical in small engine work since there is usually plenty of room to reach the fasteners. However, the size of the teeth have much to do with tool wear. In general, you should choose a wrench with heavy teeth. You can get an idea of the pitch by working the action; the larger the teeth, the greater will be the arc of handle movement between clicks.

Figure 2-2 shows two ratchet handles: a Husky ½in. drive, and a Craftsman ⅜ in. swivel-head. Swivel-head wrenches were developed to change the spark plugs on American V-8 automobile engines and have become increasingly popular for general repair work.

A *breakover bar*, or breaker, has a solid shank and a swiveled head like the ratchet. It is designed to do jobs which would quickly destroy a ratchet handle. You can use a breaker in conjunction with a cheater bar (a length of steel pipe slipped over the handle) to multiply the leverage, or even a hammer (if you're not averse to marring the finish). Even without these aids, a breaker generates tremendous torque; a strong man can outperform an electric impact-wrench with one of them. Purchase a ½ in. drive with a 24 in. handle and an adapter to bring the drive size down to ⅜ in. for the smaller sockets.

A *hammer driver* is useful, but not absolutely necessary. It consists of a heavy metal cylinder and piston. When the piston is struck with a hammer it moves the driver end left or right. These tools are excellent for removing butchered Phillips screws, since the bit is driven down by the hammer as it turns. A *speed* wrench looks like an old-fashioned automobile crank. In the ⅜ in. size the handle flexes before much torque can be generated, but as the name indicates, these wrenches are fast. They find their best use in removing and installing cylinder-head and flange bolts.

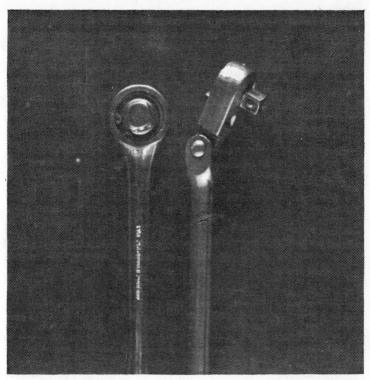

Fig 2-2 Business end of two ratched handles. The larger one is a Husky CS 43 mentioned in the text. The swivel-head is useful on riding mowers and tractors where space is a premium.

Power-impact wrenches are the province of the professional. Generally, the electric types that develop torque in ranges of 90 to 130 ft-lb are preferred. Air-driven wrenches are easier to handle than some of the electric types, but require a source of compressed air. Either type can be converted into a heavy-duty drill with a chuck available from Milwaukee Electric Tool (13135 West Lisbon Rd., Brookfield, Wis. 53005).

A *torque* wrench is indispensable. These wrenches measure the amount of torque applied to the handle.

Engine blocks, heads, flywheels, and connecting rods look as imperturbable as brass monkeys. But under compressive and inertial stresses these parts, which seem so stolid, bend and flex. The problem is more acute with aluminum, although it is present in iron as well. The fasteners—screws, bolts, nuts, pins, lockrings, and the like—not only hold the engine together as an assembly, but also provide reinforcement against working stresses.

Assume, for this discussion, that rod cap nuts have a torque specification of 10 ft-lb. This means the factory has determined that the rod cap undergoes dynamic loading equivalent to 10 ft-lb on the fasteners. At TDC, the primary shaking force is effectively canceled as far as big-end structural integrity is involved. In other words, the preload on the rod cap fasteners is balanced by the shaking force. If the rod fasteners were undertorqued, they would stretch under inertial loads. At 3000–3500 rpm, small elongations soon open the crystalline structure of the studs and fatigue cracks develop.

If, conversely, the rod fasteners are overtorqued, the journal will distort. Clinton has determined that when torque exceeding factory specification by 50% is applied to an aluminum rod assembly, it will narrow the working clearance by 0.01 inch. A worn rod might survive this kind of deformation—although wear will be concentrated on a small part of the bearing radius—but a new rod can easily fail. New clearances are on the order of 0.001 and 0.002 inch.

While the big end is the most critical rod component, and the one whose demise is announced by flying bits and pieces, correct torque is important for other parts as well. Insufficient head-bolt torque almost guarantees a blown gasket and can lead to valve problems. (Once the gasket begins to leak, the fuel mixture is made lean and chamber temperatures increase. Too much torque creates bore, valve-seat, and guide distortion. A loose flywheel nut can make starting difficult, damage the crank and hub, and shear the flywheel key. At the other extreme, the wheel may shatter from compressive loading.

Since the fasteners serve as beams and columns, the torque application sequence is very important. Assemblies tightened in a random fashion may warp. As a general rule, work from the inside outward. This rule is subject to modification for asymmetrical parts such as cylinder heads, however. Tighten in increments to distribute the load evenly as the parts come together. A cylinder head, for example, should be taken down in at least three increments.

Another reason for investing in and using a torque wrench is the nasty little problem of broken and stripped fasteners.

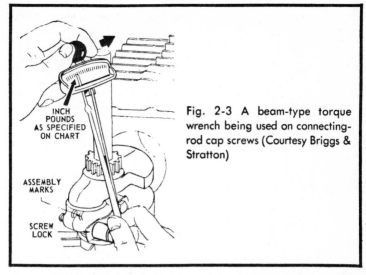

INCH POUNDS AS SPECIFIED ON CHART

ASSEMBLY MARKS

SCREW LOCK

Fig. 2-3 A beam-type torque wrench being used on connecting-rod cap screws (Courtesy Briggs & Stratton)

Those which screw into *blind* bosses create the most havoc. You can drill and back out a broken stud and retap stripped threads—but it's so much simpler not to get into trouble in the first place.

Unfortunately, the torque we apply to a fastener is only indirectly related to the tension or pull generated by the threads. Many variables intervene. Rust, caked grease, thread profile deformations, and inaccuracies in the wrench can make a travesty of our efforts to balance the stresses in the engine. One study suggests that friction accounts for 90% of the torque input at near-specification values. If frictional losses go beyond the factory-supposed value by 5%, the final tension on the fastener would only be half its intended value. To keep friction within reasonable bounds, it is necessary to keep the threads clean; do *not* use a power-driven wire wheel to do it, however.

Applying antiseize compound (available from auto supply stores) to the threads and to the contact side of the heads will also minimize friction.

Use an accurate wrench—most wrenches are designed to be accurate within 2% of their midscale value. This degree of accuracy is comparable to that obtained at the factory with automatic tools. The more expensive wrenches have micrometer dials or a cam arrangement which releases the wrench at a preset value; some make an audible click. The best choice for the average mechanic is the direct-reading *beam* wrench (Fig. 2-3). Beam wrenches are quite inexpensive, have but a single moving part, and can be calibrated by bending the needle to zero on the scale when no load is applied to the beam. You will need one in ⅜ in. drive, rated at 0–50 ft-lb for rod and head work. The larger tractor and riding-mower engines will require a larger capacity wrench for the flywheel.

Whenever using a wrench (torque or any other type), *always pull toward you.* Pushing on a wrench is an invitation to skinned knuckles.

Allen, or recessed hex-head wrenches, are necessary when dealing with drive pulleys. The Allen brand costs more than some of the others and may require a trip to a large tool supply house to find it. But you will appreciate the quality when you encounter a rusted setscrew.

Screwdrivers

There are two philosophies about screwdriver quality. Some mechanics buy cheap ones by the handful and use them for chisels and punches as well as for their intended purpose. Others buy the best and take care of them. The latter course is perhaps more satisfying and is undoubtedly cheaper in the long run. Some of the finest screwdrivers available are sold through the mail by Ames Supply (3170 Draper Drive, Fairfax, Virginia 22030). These screwdrivers come in three weights from "extra heavy" to "light." The middle or "heavy" style is best. The shanks are round and the handles are of close-grained wood. Two bit-widths are offered in this series—³/₁₆ in. for the 4 and 6 in. lengths, and ⁷/₃₂ in. for the 8 in. length. The shanks are the same diameter as the blade width, which means that the drivers can be resharpened indefinitely.

Heavier flat-blade and Phillips drivers can be purchased from tool jobbers and large hardware stores. Insist on name brands and scrutinize the tool carefully before you pay for it. A visual examination cannot tell you anything directly about the quality of the alloy or the heat treatment, but it will give you clues. Bits which are cast are already dull. Handles with sharp serrations have been designed more for shelf appeal than for comfort. And, in general, you should steer clear of "trick" drivers with built-in rachets, exotic screw holders, and other questionable refinements—the simpler, the better.

Pliers

The familiar combination pliers should stay in the kitchen drawer with the other utensils. Needle- or long-nose pliers (referring to the relative taper of the jaws) are convenient as heatsinks when soldering and as an aid to holding small parts. These pliers are not designed to transmit torque. The best in quality are reserved for the electronics trade, and are available from Allied (2400 Washington Blvd., Chicago, Ill. 60612) and other mail-order houses.

Sooner or later you are going to encounter a *snapring*. Two types—contracting and expanding—are used on lawnmower shafting. You will find them on wristpins, transmission and clutch shafts, and on several rewind starter mechanisms. Without the proper tool, the temptation is to use an icepick or a knife blade. You can usually remove the ring, but you may warp it in the process. A failed snapring means trouble. In the case of the wristpin, the engine may be destroyed; so purchase a good inside and outside snapring tool before you need it.

Vise-Grips, or locking pliers, have a poor reputation in some quarters. When used indiscriminately, the tool can be destructive. It generates a holding force of 3 tons and applies it through serrations milled on the jaw faces. Naturally, fasteners subject to such localized force take on a dog-eared appearance—the battered fastener is the hallmark of the amateur mechanic.

Use locking pliers only as a last resort, after gentler methods have failed. Of course, if the fastener is already rounded, or rusted solid to its threads, you needn't be reluctant. The worst that can happen is that the fastener breaks.

Since Vise-Grip's original patents have expired, scores of locking pliers have entered the tool market. Many—if not most—are inadequate. Cheap tools are made of pressed steel, welded and reinforced at the jaws. The force the handles generate is merely displaced through the stampings. You bend the pliers rather than squeeze the fastener. The Vise-Grip brand is probably the best choice and, figured on a cost-per-use basis, the most economical. They are available at large hardware stores and at some of the independent auto parts jobbers. Specify the 8 in. size with flat jaws. The round-jawed version is more appropriate for plumbing work.

Besides loosening stubborn fasteners, a pair of Vise-Grips serves as a portable vise, a file holder, and as a powerful clamp.

Hammers

In an ideal machinist's world, there would be no need for hammers. The parts would be mated perfectly, and where there was an interference fit, we would have a convenient arbor press. Of course, the real world is less perfect; every mechanic depends on an assortment of hammers. You will need a ball peen or machinist's hammer in the 8-ounce size. If you plan to do much work on heavy garden equipment, you can supplement the 8 oz with a 12 or 16 oz.

Using a hammer requires some skill and a little common sense. Aluminum castings are notoriously brittle and are subject to subsurface fatigue cracks. A light tap may shatter a crankcase—even if the casting doesn't shatter, it may bend. Cast iron is stronger and causes fewer surprises, but is by no means impact-proof. If it is necessary to hammer on a casting, use a wooden block to cushion the blows and spread the force over a wide area; concentrate the blows on parts of the casting which are reinforced by ribs or webbing. And know when to stop. For example, it is common practice to remove flywheels with a knocker. Although this practice can hardly be termed beneficial, it is done thousands of times a day in shops everywhere without producing obvious damage. But a really stubborn flywheel may not respond to a knocker. I once saw a motorcycle engine with the crankshaft driven through the block and the flywheel still in place!

Hammers should be kept clean and dry; a greasy hammer is a menace to the user and anyone in his vicinity. The handle should fit the head tightly and be free of obvious defects. Scrap a hammer which has started to "peel" or flatten. Hammer heads are heat-treated for surface hardness. The inner metal is quite soft to prevent fracturing—but once the integrity of the outer surface is broken, the hammer will throw off chips. The better repair shops insist that mechanics wear safety glasses.

Soft-faced mallets are convenient but not absolutely necessary. Although this is a matter of opinion, in general, a rawhide mallet is superior to the plastic or rubber varieties. Lead hammers are specialized tools, and not the sort of thing that one would run out and buy to work on lawnmowers. But a lead hammer is nice to have around; you can use it on all but the most vulnerable steel without leaving "face prints." If you want one, try a large tool house.

Punches and Chisels

Most drive trains are held together with ⅛ in. compression pins. You will need a punch to replace the gears, shafts, and wheel rollers. An 8 in. punch also serves as a backout tool for broken fasteners. The tip, in the shape of a Capehart chisel, will gouge deeper as it cuts and will give you a mechanical advantage on the fastener; it should be ground to a 45° angle. A cold chisel should also be part of your tool kit as an ultimate persuader for rusted fasteners.

Files

You will need an 8 in. *mill bastard* file for general cutting. It can be used to finely sharpen rotary lawnmower blades, flat screwdrivers, and the like. A *rattail* file is useful to enlarge holes as, for example, in the sheet metal on riding mowers. A small triangular file can be used to dress keyways, remove burrs, and to make stress riser cuts on spring steel cable.

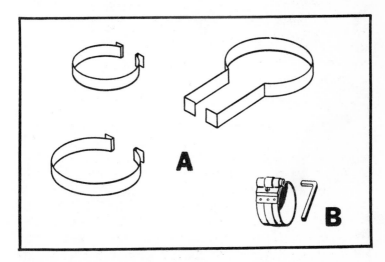

Fig. 2-4 (A) Ring compressors for 2-cycle engines (Courtesy Kohler of Kohler) (B) Ring compressor for 4-cycle engines (Courtesy Clinton Engines Corp.)

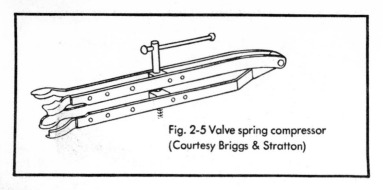

Fig. 2-5 Valve spring compressor
(Courtesy Briggs & Stratton)

Triangular files can do wonders for worn Phillips screwdriver bits.

Special Tools

Engine manufacturers stock special-order tools for their mechanics. In a few instances, these tools are absolutely essential, and they always speed the work. You will find, however, that you can substitute other tools for most factory items.

The need for special tools is most pronounced during major engine repairs. Piston rings, main bearings, and oil seals are not designed with ease of replacement in mind. You should purchase a piston-ring compressor as illustrated in Fig. 2-4B. The Clinton tool has a range of 1¾ in. to 3⅛ in. and will fit all but the largest riding mower engines. Bore sizes greater than 3 in. can be handled by an automotive ring compressor. The 2-cycle compressors shown in Fig. 2-4A open to allow the cylinder to be fitted.

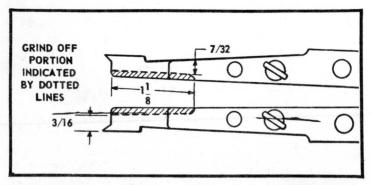

Fig. 2-6 Modification of the Briggs compressor for small-block engines—not required on late-production tools (Courtesy Briggs & Stratton)

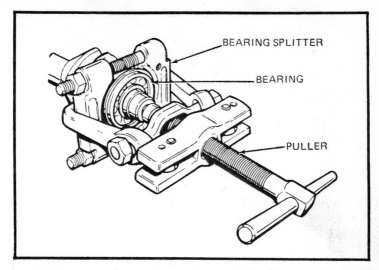

Fig. 2-7 Bearing removal with a splitter and puller (Courtesy Tecumseh Products Co.)

For valve work, you most certainly will need a lifter, although everyone seems to do their first few valve jobs with the aid of two screwdrivers. Perhaps the best lifter for small engine valves is made by Briggs & Stratton (Fig. 2-5). Some light grinding is necessary along the frame ends (Fig. 2-6) to squeeze the tool into the valve chambers of the smaller engines. Valve seat and face grinders are nice to have, but it's much cheaper to farm out this sort of work to an automotive machinist.

Ball and tapered roller main bearings must be removed from the crankshaft with a bearing splitter or puller, as shown in Fig. 2-7. Attempting to drive the bearing off will bend the crankshaft. In most cases the dealer or an automotive machine shop will have the appropriate tool. Bushing main bearings can be renewed by the dealer. The tool cost is prohibitive for the average individual.

Flywheel removal is a delicate operation. Some manufacturers insist that their wheels be removed with a puller (Fig. 2-8). The puller illustrated has self-threading screws which fit into recesses cast into the hub. Other manufacturers allow a conventional gear puller to be used on their flywheel rims. Others discard the puller idea altogether, and supply *knockers*.

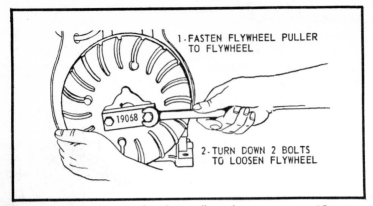

Fig. 2-8 Flywheel removal with a puller—the correct way (Courtesy Briggs & Stratton)

Knockers (Fig. 2-9) are heavy bolts which thread over the crankshaft end and are hand-tightened and struck with a hammer. The crankshaft telescopes a few thousandths of an inch and releases its grip on the flywheel. Some of those who oppose knockers do so for fear that the shaft may take a permanent set at the rod journal. Others are uptight about the vibration; it has been known to damage the flywheel magnets by rearranging the molecular structure (demagnetization).

Clinton offers knockers in two sizes, and Tecumseh offers one with left-hand threads. These three knockers will service all lawnmower engines except those Briggs & Stratton engines with impulse and rewind starters. The crankshafts have a tapered extension, chamfered at the end. It is almost as if B&S had built the crankshaft to be knockerproof. However, just as no lock is secure from picking, no crankshaft is safe from knocking. Begin with a bar of Monel steel, approximately 1 inch in diameter and 4 inches long. With a lathe, drill a ½ in. hole in the end to a depth of 3 inches. Grind the drill bit 90° to the shank instead of the usual 57°. Finish the hole, leaving the end square. The last operation is necessary to prevent the knocker from binding on the beveled edge of the crankshaft.

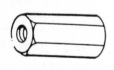

Fig. 2-9 Flywheel knocker—not recommended by most manufacturers.

Miscellaneous Supplies

Degreasing is one of the most time-consuming aspects of the mechanic's trade. Commercial operations can afford to invest in air-driven spray machines and in steam generators. The amateur must rely on muscle. There are products on the market which will speed the job. One is known, somewhat inelegantly, as Gunk. Available in spray cans and quart tins, it is mixed with kerosene before use. Brush it on and hose it off. Gunk is a strong base (alkali) and acts upon grease to convert it into soap. In the process, it cuts aluminum oxide and gives the castings a like-new look. The only drawbacks are that Gunk attacks paint, fading the darker colors, and must be rinsed with water. If used on internal parts, care must be exercised to prevent rust.

Kerosene or Varsol is safer to use on disassembled components and will dissolve all but the most stubborn grease and oil accumulations. Diesel oil (No. 2 heating oil) may also be used and has such a low flash point that it can be considered combustion-proof. Of course, gasoline should not, under any circumstances, be used as a solvent. Some mechanics keep a tin of paint remover handy for cleaning carbon from piston ring grooves. Allow pistons to soak for several days.

Emery cloth stripping is sold by the roll or foot. Specify a medium-grit, wet-or-dry type. Dry, it is used to remove rust and scale from exposed crankshaft ends before engine teardown; wetted with oil, strip cloth is used to burnish small imperfections from bearing journals. When mounted on a rod (the end slotted to hold the cloth in place) it converts an ordinary drill into a poor man's inside miller. A recently introduced 3M abrasive has the appearance of screen wire. The holes in the mesh prevent clogging and extend the life of the abrasive.

Some mechanics swear by gasket cement. Others consider its use a sign of failing confidence. Two clean and reasonably parallel surfaces can be joined with a gasket without the aid of ticky-tacky. Nevertheless, cement should be used behind the seals at the metal-to-metal interface, and on gasketless joints. It should not be used to "repair" a gasket or to replace one. Permatex claims that their new gel obviates the need for all gaskets except those subject to exhaust-gas heat.

One of the handiest products, available from electronics supply houses, is alpha-cyanoacrylate cement. A superstrong contact cement, it's widely used in industry to *replace* conventional fasteners. When applied to clean surfaces it has a tensile strength of 5000 lb/sq in. You could put an engine together with it if you were not concerned about taking it apart again. Cyanoacrylate can be used to anchor stripped studs in their bosses, bridge pinhole casting defects and fatigue cracks, secure valve guides, and to do hundreds of other jobs which once bordered on the impossible. You can even "weld" cast decks with it. Although one might be seduced into attempting almost *any* repair with this cement, some words of caution: Do not get any on your fingers—if you do, dissolve the bond with fingernail polish remover or acetone.

Chapter 3

Sighting in on the Problem

Diagnosis is the most challenging aspect of small engine repair. After you have overhauled a score or so of engines, the learning curve flattens—that is, you learn less and less with each job. The only compensation is that you get faster. But troubleshooting is always a fresh challenge. No one, not even mechanics who have worked in the trade for years, knows everything about troubleshooting.

Troubleshooting is a game of probabilities. The engine deals your hand, and counters each play you make. It takes skill to tilt the probabilities in your favor, and to make the right guesses. Here is general procedure to give you an edge in this game:

1) Isolate the problem to one system; i.e., ignition, compression, fuel, charging, or propulsion.

2) Use whatever information you have gleaned from experience, theory, or common sense, to get a fix on the most likely component; we know, for example, that in the event of ignition failure, the points are the most likely suspect. The condenser is next, followed by the coil.

3) Considered your initial diagnosis tentative—be ready to change it as new information comes in; a mechanic can waste a great deal of time and energy by "overinvesting" in his initial diagnosis.

4) Choose the simpler alternative between two suspected causes for a malady; most failures have a very simple origin.

5) Think in terms of linking problems; e.g., the engine will not start because the points are out of adjustment and the carburetor is partially clogged. Neither one would be enough to put the quietus on the engine, but both together

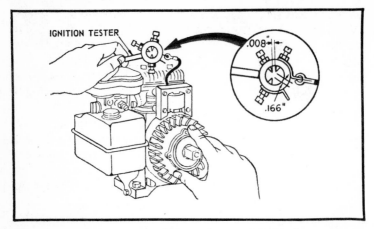

Fig. 3-1 A spark output tester. If the arc jumps a 0.166 inch gap you can assume that the magneto is capable of starting the engine (Courtesy Briggs & Stratton)

can run you in circles. If you suspect linkages—and you can have four, five, or six in an old engine or one that has been tinkered with—give up trying to pinpoint the exact problem. Overhaul each system.

The engine should run if it has compression, fuel (in the right proportion to air), and a spark (at the right time). This motto should be engraved in the mind of every mechanic. In the remainder of this chapter, we will discuss typical malfunctions and troubleshooting drills.

Refusal to Start

Some years ago Clinton discovered that 87% of small engine repairs involve the magneto ignition system. Magnetos have improved since then, but are still the most failure-prone component. Begin every troubleshooting sequence with a spark output test. Automobile mechanics have access to instruments (Fig. 3-1) to test this and other systems. Small-engine mechanics are not so fortunate and must use less exact methods. But this is what makes the game interesting.

Output is estimated by the ability of the spark to jump an air gap and by spark appearance. Figure 3-2 shows a test plug being used on a Clinton engine. The plug is an 18mm Ford type with the ground electrode bent to provide a $5/32$ to $3/16$ inch air gap. Ground the shell to the block or shroud as shown. If you don't, ignition voltage will pass through you. While you won't be electrocuted, your experience will certainly be unpleasant. (Many people have been seriously injured by cuts sustained in sudden reaction to a minor electrical shock.) The voltage produced by the new CDI systems is said by the manufacturers to be lethal. When handling high-voltage cables, use a pair of cable pliers of the type sold in auto supply stores.

The spark should jump the gap at cranking speed (90 rpm). It should be sky-blue and fat. A really good magneto will pump enough voltage to produce an audible crack as the spark burns across the air gap. White or red sparks indicate dirty points and possible condenser trouble. Thin, spindly sparks indicate low voltage.

Assuming that the spark appears healthy, replace the spark plug with one known to be good. We only have one on a lawnmower and a slight imperfection can prevent starting. Formerly, I would have simply said substitute a new one; but since the fuel shortage, new spark plugs have become somewhat unreliable. One out of fifty or so will not fire under compression. Whether this has to do with chemical changes in the fuel or poor manufacturing procedures is unknown.

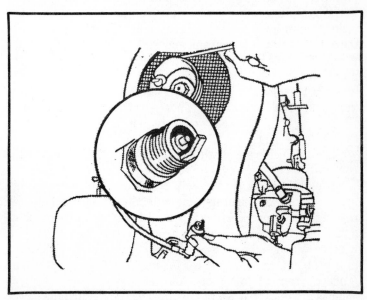

Fig. 3-2 Checking the spark. The plug shell must be pressed firmly against the shrouding or block. Failure to make good contact will put you in the ground return (Courtesy Clinton Engines Corp.)

Fig. 3-3 Checking compression (Courtesy Clinton Engines Corp.)

Examine the old plug for evidence of wet fouling. If the plug is damp, the engine is flooded with raw gasoline. You can dry the cylinder with compressed air or by cranking with the plug out. It can take 15 to 20 pulls to evaporate the fuel. Two-cycle flooding may also involve the crankcase. Since crankcase drain plugs are as scarce as snowshoes in Tahiti, the only alternative is to crank and wait. Using compressed air is not recommended because of the danger of distorting the reeds.

A hot engine can be flooded but still have a dry plug, indicating that the fuel has vaporized. Gasoline ceases to be flammable when its vapor concentration in air reaches more than 6% by weight.

Lack of fuel in the chamber means a problem in either the compression sector or in the fuel delivery system. A compression test is not entirely reliable since many factors intervene between average engine compression and the psi reading you get on the gage (Fig. 3-3). Ideally, you should keep a running check on your engine from the time of purchase. A wide variation in the compression curve would then alert you to a problem. In very general terms, 4-cycle engines should produce from 65 to 80 psi. Large, slow-turning cast-iron engines are clustered low on the pressure scale, and the newer aluminum-block designs are in the 80 psi range. Two-cycles are generally low-compression devices producing 60 to 70 psi.

In order to obtain uniform readings, the test procedure must be standardized. Ground the ignition and open the throttle and choke plates. We want as little restriction as possible on the carburetor. Crank, and take the reading on the fourth pulse.

Too little compression—few engines can be persuaded to start at 40 psi or less—means ring or valve leakage. You can localize the problem by squirting 2−3 cc of oil into the chamber. If the compression fails to jump, the valves are at fault. Do not make this test if you expect to use the engine immediately. Oil in the chamber is a particularly objectionable form of flooding and may entail flushing with a solvent.

Too much compression—and this is where you wish you had kept a record—means an accumulation of carbon in the chamber. In extreme cases, carbon deposits can raise the compression ratio to the threshold of detonation.

Two-cycle engines develop crankcase compression a few pounds over atmospheric pressure at cranking speed. No convenient way of measuring crankcase compression has been devised. Case, seal, and reed valve integrity can be determined with an air pump after the engine is dismantled; but this procedure is tedious and time-consuming. Mechanics who specialize in these engines estimate crankcase compression as it translates into resistance on the starting cord. Fortunately, seal and reed valve problems are rare.

Fuel system difficulties which would prevent starting usually can be traced to the choke. Lawnmower chokes must be fully closed during cranking (Fig. 3-4). With age, the

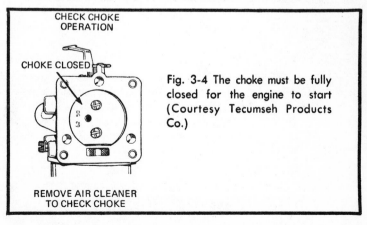

CHECK CHOKE OPERATION

CHOKE CLOSED

Fig. 3-4 The choke must be fully closed for the engine to start (Courtesy Tecumseh Products Co.)

REMOVE AIR CLEANER TO CHECK CHOKE

BEND HERE

Fig. 3-5 The Bowden (control) cable can be moved at its carburetor mounting clip or, in some cases, the choke-actuating linkage may be bent to obtain full closure (Courtesy Briggs & Stratton)

Bowden cable, or linkage, may flex enough to allow air to pass around the choke valve (Fig. 3-5). The Briggs automatic choke should be closed regardless of engine temperature.

Spit-back through the carburetor can mean the ignition timing is off—very rare, and almost always the repairman's fault. A leaking intake valve can be another cause of this misbehavior—check compression with a gage. Spit-back is often symptomatic of a loosening in the flywheel, rotary mower blade, or blade belt.

Lack of Power

This is a difficult problem to troubleshoot, since it is in part a subjective reaction by the owner and since so many

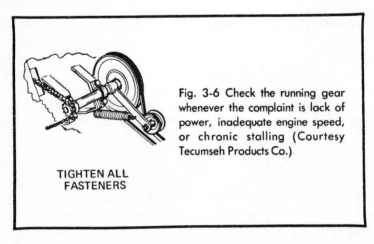

TIGHTEN ALL FASTENERS

Fig. 3-6 Check the running gear whenever the complaint is lack of power, inadequate engine speed, or chronic stalling (Courtesy Tecumseh Products Co.)

mechanical factors can be involved. First decide if the machine is being operated within its load limits. According to research done by Briggs & Stratton, the average lawn requires about 1 hp from a rotary mower. But this factor can multiply in heavy grass.

Check the running gear (Fig. 3-6) on self-propelled models. A slipping belt or tight bearing can give the illusion of inadequate power. Blade clutches on riding lawnmowers are also a prime suspect.

Remove the spark plug and examine the firing tip as explained in Chapter 5. The condition of the tip will give you clues about mixture strength and, possibly, about timing. Replace the plug with a good one and test the mower under load. Check the compression pressure. If you can't uncover any clues—the engine may be simply worn out; 2-cycles sometimes develop arteriosclerosis of the exhaust ports or the muffler. Be sure to examine the air filter, carburetor mounts, and main jet adjustment.

Engine Misses

Although the carburetor shouldn't be at fault, it usually is—particularly when the owner has changed the adjustments. Set as outlined in the following chapter and replace the spark plug. If missing persists, investigate the magneto. Missing can be caused by any magneto component and makes coil trouble more likely than it normally would be. Malfunctioning in the carburetor internals is the least likely trouble source, if we discount really abstruse items like reed valve flutter and stratified layers of water in the fuel tank.

Engine Won't Idle

The obvious choice here is the low-speed circuit in the carburetor, followed (in order of likely culprits) by small air leaks and incorrect carburetor assembly. If the engine has remained dismantled since the last time it ran, check the valve timing as a last resort. Idling difficulties also can originate in improper magneto timing or point gap. However, these problems usually surface as a refusal to start.

Flat Spots in Acceleration

Suspect the carburetor for this one. Small engine carburetors are not the most sophisticated instruments and often have a lean transition from idle to cruise. It usually can be masked by cracking the main jet wider.

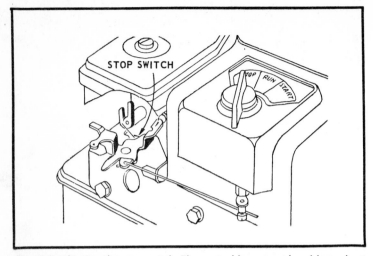

Fig. 3-7 Adjusting the stop switch. The control lever must be able to short the primary at one end of travel, and choke the carburetor at the other. Generally you can obtain both functions by adjusting the Bowden cable. Otherwise, bend the linkage (Courtesy Briggs & Stratton)

Engine Runs Intermittently

Check the fuel system first, because it is most accessible. Look for a plugged gas cap or float bowl vent. Next, turn to the ignition. Replace the condenser, then the coil. Substitution is the only choice, since few shops can test electrical components while installed.

Refusal to Shut Down

Check the magneto shorting switch at the carburetor. In most instances, the remote linkage will be found to be biased in favor of the choke. Adjust for contact with the switch, without losing choke control (Fig. 3-7). Run-on, or *dieseling*, is identified by uneven idle and is caused by extreme overheating. Check the timing and the oil level.

Seizure

To verify that the engine is in fact immobile, and the problem does not lie in the blade drive (those models with clutches), wheel drive, or starter mechanism, turn the crankshaft by hand.

Certain safety precautions must be observed to prevent accidental starting. Many mechanics have been hurt when a "dead" engine suddenly came to life. Ground the ignition primary (if an ignition switch is provided) and, as additional insurance, ground the high-voltage lead firmly to the block. You do not want it to fall back into contact with the spark plug; some mechanics remove the plug to avoid this hazard.

Before operating, stand your mower against a wall (cylinder head up on 4-cycles to prevent oil flooding). Remove the wet air filter on those engines which have these devices—otherwise, oil will spill into the carburetor throat. Block the wheels.

Try to rotate the blade by hand; if it does not turn, check the clutch mechanism. A jammed clutch simulates engine seizure. Check the propulsion mechanism; remove the shroud to be sure the starter—mechanical or electric—is not binding on the flywheel. If the crank still refuses to turn, the engine is at fault.

Three conditions can cause seizure. An engine which has been in storage several months may have its rings rusted to the bore. If this is the case, loosen with penetrating oil and by applying pressure to the crank. Remember, the blade is sharp—wear gloves. Horizontal crankshaft engines can be freed at the crank stub with a pipe wrench. Be gentle, since the crankshaft bends easily.

Hydraulic lockup is rare, but can happen when an engine is tilted so that oil flows by the rings to the chamber. This requires flushing with a solvent. It is not recommended that you use gasoline or other solvents with a low flash point.

The most common cause of seizure is, unfortunately, the most expensive to correct. Overheating causes the bearing surfaces to expand and fuse together. In 4-cycle engines, crankpins are usually affected first, although the main bearings may also be involved. Pistons seize first in 2-cycles, followed by crankpin failure. Generally, overheating can be traced to insufficient lubrication. You will find little or no oil in the sump, or raw, unmixed gasoline in 2-cycle tanks.

The engine should be torn down and inspected, whether or not the piston can be freed. However, a new engine can sometimes be put back into service without inspection. This practice is common in dealer's shops. Customers purchase a mower and fail to put oil in the sump or neglect to premix the fuel—the engine runs a few minutes and seizes. Since most mowers are purchased on credit, and since this is not a problem which the factory will warrant, the mechanic knocks the crank free with a two-by-four and returns it to the customer.

Excessive Smoke

Sooty, black smoke is infallible evidence of an overrich mixture. Blue smoke is a function of the percentage of oil in the fuel. All 4-cycle engines burn some oil; if they didn't, the upper ring would run dry. Small engines burn more oil than the typical automobile on a pounds per horsepower-hour basis. Under transient conditions, these engines will emit puffs of blue smoke.

Excessive and chronic smoking is usually a sign of worn rings, which have lost their temper from overheating. A compression test can verify ring condition. Other causes are an overfilled sump or filter reservoir. The mower must be level during servicing. A clogged crankcase breather may pressurize the sump and force oil past the rings.

Two-cycle machines produce smoke as a matter of course—but great clouds of the stuff can be traced to an oil/fuel mix that is too rich or to stratification of the oil in the tank. Premixed fuel should be stored in its own container and agitated before use.

Vibration

Vibration becomes intrusive when the operator's hands go to sleep on the bars, or when the motor shimmers as if it were seen through air rising from a heated surface.

Poorly designed lawn equipment can go into resonance with primary or secondary forces. Instead of the machine acting as an anvil to absorb these impulses, it becomes a sounding board. There is little that the owner can do; although insulating the handles with rubber snubbers may be worth the effort.

Increases in vibration to a level above the norm are serious. It is not unknown for a small engine to literally shake itself to pieces.

The power takeoff end of the crank is the most likely source of unbalance. Crank stubs on horizontal-shaft engines can bend, but the phenomenon is rare. Because the crankshaft turns a small pulley, out-of-balance forces are small and, if we are willing to tolerate reduced main bearing life, can be ignored.

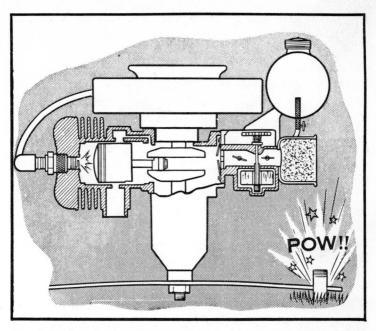

Fig. 3-8 When this happens, something has to give. Damage may be localized at the blade or, in the worst case, involve all rotating and support components, including the block and flywheel (Courtesy OMC)

Direct-drive rotary and riding mowers are another matter. If the blade strikes an immovable object, something is going to give (Fig. 3-8). Rough calculations (not taking into account the rotating mass of the flywheel and crankshaft) indicate that a 2 ft blade spinning at 3000 rpm has a kinetic energy of 12,500 ft-lb. Were this energy released, it would lift a 12,500 lb mass 1 foot in the air. Forces of this magnitude are easily capable of bending the blade, adapter, crankshaft, and engine block. In practice, some of the force is bypassed through the adapter shear pins or the clutch. But crankshaft deflection is not unusual.

To determine if the crankshaft is bent, raise the mower and disconnect the ignition as outlined previously. Sight along one blade tip to a fixed reference point on the deck. Rotate the blade 180°. The other tip should be within a fraction of an inch of the first. If it tracks significantly higher or lower, remove the blade and take a direct sight on the crankshaft center. Have someone spin the engine; wobble means that the crank is bent and must be replaced.

Fig 3-9 Decks can be welded in most instances (Courtesy OMC)

Improper sharpening can also put a blade out of balance. Purchase a static balancer from your local lawnmower parts supply house. The metal types seem to be more accurate than those of plastic.

Check the shroud and mounting bolts. If you can see the engine tremble, it is advisable to remove it from the deck and look for cracks around the mounting holes. (Steel decks can be welded or brazed—Fig. 3-9.) It is possible to arcweld all cast aluminum and some magnesium decks.

Noise

Where there's vibration, noise cannot be far behind. Make the checks outlined above, giving special attention to the shroud and engine mounting fasteners.

Mufflers work in a mist of water and acid. Cast aluminum mufflers (used on many 2-cycles), seem to thrive in this environment; but steel mufflers should be routinely replaced. Use a factory original to keep back-pressure and exhaust valve life within normal limits.

Mechanical noises can be divided into two groups. Some consist of assorted bangs, raps, and clangs. Loose big-end rod bearings knock when the engine is coasting down. If the piston were under sustained load, no connecting rod cap would be needed. Main bearing noise is duller sounding than that generated at the crankpin, and can be heard at all speeds. Tappet noise is not a matter of urgent concern—unless we are interested in maximum cylinder filling. It can be heard as a series of light raps.

The other category consists of rips, squeals, shrieks, and groans. These noises emanate from overheated bearings or from interferences between rotating parts.

One of the most hair-raising sounds is caused by a misaligned starter drive in partial engagement with the flywheel hub. To squelch the racket, loosen and then retighten the shroud bolts. Sometimes a drop of oil will silence the idler bearing on Briggs & Stratton rewind starters; but be careful not to wet the nylon cord. Noise from the engine proper can only be diagnosed by full teardown and careful visual inspection.

Wheel rumble develops as the machine ages. The wheels go into sympathetic vibration with primary and secondary forces. It is annoying, but can be considered harmless. The cure is to replace the factory-original wheels with industrial types sold for warehouse trucks.

ENGINE TROUBLESHOOTING CHARTS
(Courtesy Lauson Engines)

2-Cycle Engines.

Cause	Remedy
ENGINE FAILS TO START OR STARTS WITH DIFFICULTY	
Too much oil in fuel	Drain and replace in proper proportions.
ENGINE LACKS POWER	
Ports carboned	Clean ports.
ENGINE OVERHEATS	
Lack of lubrication	Check fuel oil mix.

4-Cycle Engines.

Cause	Remedy
ENGINE FAILS TO START OR STARTS WITH DIFFICULTY	
No fuel in tank	Fill tank with clean, fresh fuel.
Shut-off valve closed	Open valve.
Obstructed fuel line	Clean fuel screen and line. If necessary, remove and clean carburetor.
Tank cap vent obstructed	Open vent in fuel tank cap.
Water in fuel	Drain tank. Clean carburetor and fuel lines. Dry spark plug points. Fill tank with clean, fresh fuel.
Engine over-choked	Close fuel shut-off and pull starter until engine starts. Reopen fuel shut-off for normal fuel flow.
Improper carburetor adjustment	Adjust carburetor.
Loose or defective magneto wiring	Check magneto wiring for shorts or grounds; repair if necessary.
Faulty magneto	Check timing, point gap, and, if necessary, overhaul magneto.
Spark plug fouled	Clean and regap spark plug.
Spark plug porcelain cracked	Replace spark plug.
Poor compression	Overhaul engine.

Cause	Remedy
ENGINE KNOCKS	
Carbon in combustion chamber	Remove cylinder head or cylinder and clean carbon from head and piston.
Loose or worn connecting rod	Replace connecting rod.
Loose flywheel	Check flywheel key and keyway; replace parts if necessary. Tighten flywheel nut to proper torque.
Worn cylinder	Replace cylinder.
Improper magneto timing	Time magneto.
ENGINE MISSES UNDER LOAD	
Spark plug fouled	Clean and regap spark plug.
Spark plug porcelain cracked	Replace spark plug.
Improper spark plug gap	Regap spark plug.
Pitted magneto breaker points	Replace pitted breaker points.
Magneto breaker arm sluggish	Clean and lubricate breaker point arm.
Faulty condenser (except on Tecumseh Magneto)	Check condenser on a tester, replace if defective.
Improper carburetor adjustment	Adjust carburetor.
Improper valve clearance	Adjust valve clearance.
Weak valve spring	Replace valve spring.
ENGINE VIBRATES EXCESSIVELY	
Engine not securely mounted	Tighten loose mounting bolts.
Bent crankshaft	Replace crankshaft.
Associated equipment out of balance	Check associated equipment.
BREATHER PASSING OIL	
Engine speed too fast	Use tachometer to adjust correct RPM.
Loose oil fill cap or gasket damaged or missing	Install new ring gasket under cap and tighten securely.
Oil level too high	Check oil level.
Breather mechanism damaged	Check reed plate and assembly and replace complete unit.
Breather mechanism dirty	Clean thoroughly in solvent. Use new gaskets when reinstalling unit.
Drain hole in breather box clogged	Clean hole with wire to allow oil to return to crankcase.
Piston ring end gaps aligned	Rotate end gaps so as to be staggered 90° apart.
Breather mechanism installed upside down	Small oil drain holes must be down to drain oil from mechanism.
Breather mechanism loose or gaskets leaking	Install new gaskets and tighten securely.

Cause	Remedy
Damaged or worn oil seals on end of crankshaft	Replace seals.
Rings not seated properly	Check for worn or out of round cylinder. Replace rings. Break in new rings with engine working under a varying load. Rings must be seated under high compression or in other words under varied load conditions.
Cylinder cover gasket leaking	Replace cover gasket.

ENGINE LACKS POWER

Cause	Remedy
Choke partially closed	Open choke.
Improper carburetor adjustment	Adjust carburetor.
Magneto improperly timed	Time magneto.
Worn rings	Replace rings.
Lack of lubrication	Fill crankcase to the proper level.
Air cleaner fouled	Clean air cleaner.
Valves leaking	Grind valves.

ENGINE OVERHEATS

Cause	Remedy
Engine improperly timed	Time engine.
Carburetor improperly adjusted	Adjust carburetor.
Air flow obstructed	Remove any obstructions from air passages in shrouds.
Cooling fins clogged	Clean cooling fins.
Excessive load on engine	Check operation of associated equipment. Reduce excessive load.
Carbon in combustion chamber	Remove cylinder head or cylinder and clean carbon from head and piston.
Lack of lubrication	Fill crankcase to proper level.

ENGINE SURGES OR RUNS UNEVENLY

Cause	Remedy
Fuel tank cap vent hole clogged	Open vent hole.
Governor parts sticking or binding.	Clean, and if necessary repair governor parts.
Carburetor throttle linkage or throttle shaft and/or butterfly binding or sticking.	Clean, lubricate, or adjust linkage and deburr throttle shaft or butterfly.

Chapter 4

Lighting the Fire

Most small engine repairs involve the ignition system; and ignition malfunctions sometimes mimic failures in other systems. Symptoms such as overheating, carburetor spit-back, missing, four-stroking, and loss of power are often the fault of the ignition system. Whatever the complaint, a wise mechanic will first change the spark plug and, unless he is already sure the problem lies elsewhere, will remove the flywheel for a look at the points.

ELECTRICITY AND MAGNETISM

The magneto is based on a discovery made by Michael Faraday over a century ago. Faraday stumbled onto one of the important interrelationships between magnetism and electricity that forms the basis for all modern ignition systems. He connected two ends of a length of copper wire to a sensitive meter and placed the wire between the poles of a horseshoe magnet. When he moved the wire so that it cut across the lines of force emanating between the magnetic poles, current was generated. The direction of flow was determined by the direction of movement relative to the magnetic field. Current intensity was found to be dependent upon the strength of the field and the speed at which the wire was moved across it (Fig. 4-1).

The converse situation causes reverse results; when current flows through a conductor, it generates a magnetic field at right angles to the path of flow. The polarity of the field depends upon the direction the current moves in the conductor. Figure 4-2 illustrates this with a compass needle. Since like poles repel and unlike poles attract, the needle—itself a magnet—pivots in response to the polarity of the larger field. The strength of the field is a function of the amount of

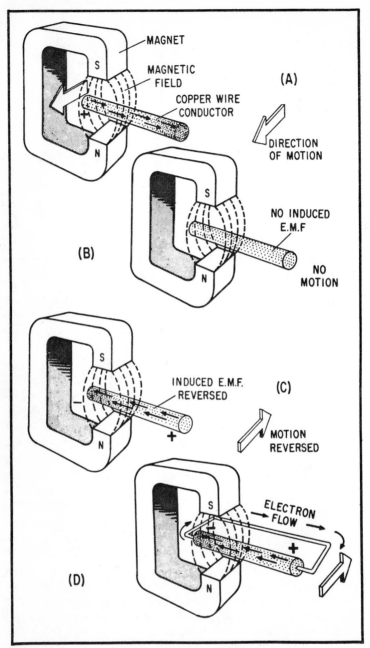

Fig. 4-1 Voltage produced by moving a conductor through a magnetic field.

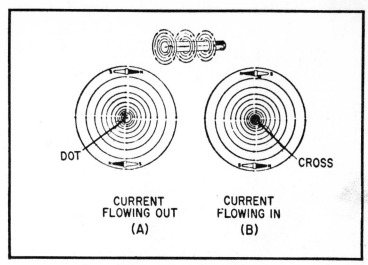

Fig. 4-2 Magnetic field around a conductor.

electrons that appear to flow through the wire. Current intensity is described in terms of amperes, or amps for short. One ampere represents 6.28×10^{18} electrons flowing past a given point in one second.

The magnetic "halos" around current-carrying circuits can affect watches, meters, and other delicate machines by magnetizing their parts. Lines of magnetic force can penetrate every substance known, but they can be directed by soft iron. Iron is *permeable* and attracts the field in a way analogous to the effect a lens has on light waves. Antimagnetic watches are protected from magnetic fields by a sheath of iron (Fig. 4-3).

A magnetic field generated by a given current flow can be intensified by winding the conductor over itself in the form of a coil. The layered magnetic fields produced reinforce each other. Further intensification can be gained by inserting an iron core into the coil. The iron gathers the lines of force and gives them direction (Fig. 4-4), and becomes an *electromagnet*. Electromagnets are temporary magnets and lose their magnetic properties as soon as the switch in their current-supplying circuit is opened. Permanent magnets operate naturally without current and retain their properties over long periods.

The magnets imbedded in the flywheels of lawnmower engines are either made of alnico (an alloy of aluminum, nickel and copper), or of a recently developed

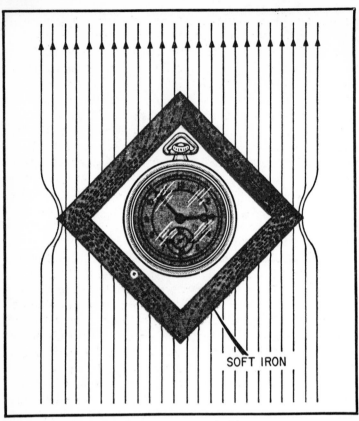

SOFT IRON

Fig. 4-3 Magnetic shield.

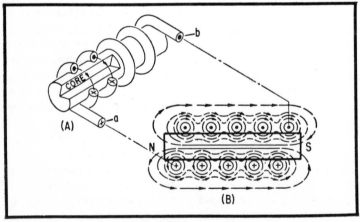

Fig. 4-4 Electromagnet.

ceramic-and-metal material. Barring accident, they remain magnetized for years—well beyond the life of the machine.

Electric current is the movement of electrons in a conductor. Matter is said to be composed of vast stretches of empty space interspersed with tiny particle systems known as atoms. Each atom has a nucleus composed of protons (positively charged bits of energy) and may have neutral particles as well. The latter, called neutrons, give mass to the atom. One or more *electrons* (with negative charges) spin in orbit around the atomic nucleus. Since unlike charges attract (negative to positive), the electrons would come crashing down into the nucleus if it were not for the centrifugal force created by their orbital velocities. Electron velocity and orbital diameter exactly match the attractive pull of the nucleus, creating a perfect system balance.

When current is applied the atomic system is flooded with free (nonorbital) electrons. The negative pole of a battery of generator has a surplus of electrons. The other terminal, the positive pole, has a scarcity of electrons. In order to equalize the charges, the surplus electrons are "willing" to travel to the positive pole through an external circuit. These electrons collide with those in orbit in the conductor. Since like charges repel (e.g., negative against negative) the orbiting electrons and displaced and jostle each other down the length of the conductor where they displace other electrons in orbit. This game, similar to musical chairs, continues until both poles of the current source have equal potential.

Of course, there is resistance to electron displacement. It varies with the atomic structure of the conductor. Some materials such as glass, dry air, most plastics, and dry wood have a very high resistance and are known as *insulators*. Conductors include silver, copper, aluminum, steel, and most metals. No conductor is perfect; some resistance is always present which converts part of the input energy of the electrons into heat. And as the conductor is heated, resistance becomes higher.

Both conductors and insulators are necessary elements in practical circuits. The conductor is the pathway and the insulator is the guard rail which keeps the electrons from taking a more "convenient" route to the positive pole. Insulator breakdown usually produces a *short circuit* in which current bypasses the conductor and whatever loads (energy-consuming working devices) are attached to it.

In general, the heating effect of current flow is a nuisance. But it becomes useful in lamps—the filament has high resistance and glows white-hot. Soldering guns, ignition glow plugs for diesel and model airplane engines, and thousands of other devices operate by current-produced heat. A few of the larger lawnmower engines employ an electric choke which opens in response to heat. The spark plug is a special case; it ignites the fuel/air mixture in the presence of heat generated by ionization of air at the plug gap.

Very high resistances virtually block current flow. If the resistance is introduced in an accidental manner, we describe the situation as an open circuit. Intentional high resistances are provided by the air gap between switch contacts, and give us control over the circuit.

Substances which have characteristics between conductors and insulators fall into a class known as semiconductors. Some semiconductor devices (like silicon diodes), have high resistance to current moving in one direction but low resistance to current flowing in the other direction. Switching transistors conduct when directed to do so by a second voltage source. They are solid-state switches, with no moving parts. You will encounter semiconductors on some of the more sophisticated lawnmower charging and ignition systems.

A circuit path must be complete and uninterrupted to be operational. If we use wire—one to the load and the other to the positive pole of the voltage source—the circuit is described as a two-wire system. You will find two-wire systems used on some riding mower headlamps and on various charging systems. But wherever possible a single wire or grounded system is employed. The load input comes by way of the engine block or the frame of the machine. The negative pole of the battery is connected (grounded) to the engine as is each load. A single insulated wire conducts the electrons back to the positive pole. The engine and frame are not insulated, but are only "hot" with respect to the positive pole of the battery. Of course, if an uninsulated portion of the positive wire touches metal, the circuit will short.

Grounded or single-wire circuitry reduces the weight, complexity, and cost of an electrical system. (Copper, the most common wire material, has very nearly taken on the status of a precious metal.) Servicing is the only disadvantage in this kind of system. Be very careful to inspect the wiring for frayed insulation and routinely scrape the ground connections. Whenever current flows between two dissimilar metals in damp air or any other conductive medium, the metals undergo a chemical change—electrolysis. The process is exceedingly complex, but it amounts to a progressive increase in resistance across the connection. Paint, oil, and rust have the same ability to block current.

Series and Parallel Circuits

A series circuit is one with a single electron path between the load and the source. In Fig. 4-5A, every electron which leaves the negative pole must pass through the lamp on its way to the positive side of the battery. A parallel circuit has two or more paths for electron flow. Figure 4-5B shows a simple parallel circuit with a second lamp "shunted" across the first.

From a technician's point of view, the major distinction between series and parallel circuits involves continuity. An opening anywhere in a series circuit will disable the complete circuit and all loads working from it. The tube filaments on ac-dc radios are wired in series; should one filament open, all

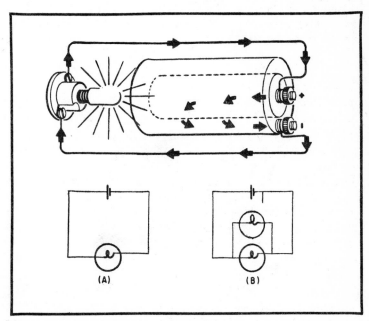

Fig. 4-5 A battery and lamp connected in series (A). Notice that there is only one path for the electrons to follow. At B, a second lamp has been connected in parallel with the first.

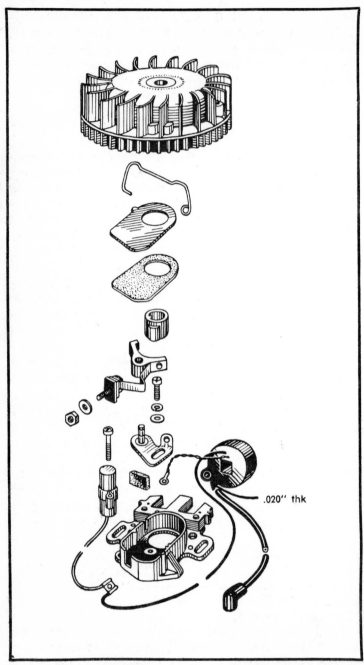

.020" thk

Fig. 4-6 Flywheel magneto (Courtesy R.E. Phelon Co.)

the tubes will go cold. Of course this characteristic is very useful when we want to be able to switch power on and off. Switches are always in series with the loads they control. Since a parallel circuit offers multiple routes for electron flow, an open in one branch will only affect the loads on that branch.

Magnetos

Most lawnmowers are sparked by flywheel magnetos. Figure 4-6 shows the parts arrangement of a magneto widely used on small gasoline engines. The flywheel carries magnets imbedded in its inner rim. The movable point shown immediately below and to the left of the cam pivots on an axle which is part of the stationary point assembly. The cylinderical object on the right is the ignition coil. One lead joins the spark plug cable; another smaller lead is grounded. The coil mounts on the center leg of the laminated iron core. The casting which locates the core is known as the stator. In this particular application, the stator has elongated mounting lugs so that it can be rotated a few degrees relative to the crankshaft—this provision enables a mechanic to time the engine.

The coil is the central element in the ignition system; it converts magnetic lines of force into electron movement through two windings which share a common ground but are electrically separate. The *primary* winding consists of approximately 200 turns of heavy wire grounded at the stator on one end and connected to the "hot" or movable point on the other. The fixed point is grounded. When the points close, the primary circuit is completed to ground. The condenser is wired in parallel with the points. It momentarily stores electrons which would otherwise arc across the point contacts when they open. Figure 4-7 shows the wiring diagram for the primary circuit. The symbols ϕ and $\pm$ mean ground; $\div$ represents the condenser, and $\equiv$ the points. The small letter is the point gap. Edge distance, which we will discuss later, is indicated at a. The secondary winding consists of several thousand turns of fine wire wound over the primary. It is grounded at the stator and at the spark plug "surround" electrode.

The metamorphosis of magnetism into electricity in an ignition coil is quite complex. Figures 4-8 and 4-9 provide a general frame of reference which, although incomplete, is

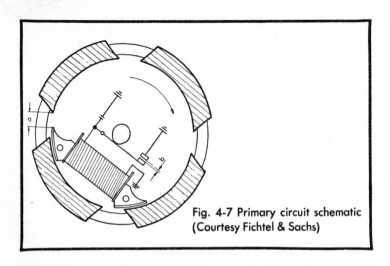

Fig. 4-7 Primary circuit schematic
(Courtesy Fichtel & Sachs)

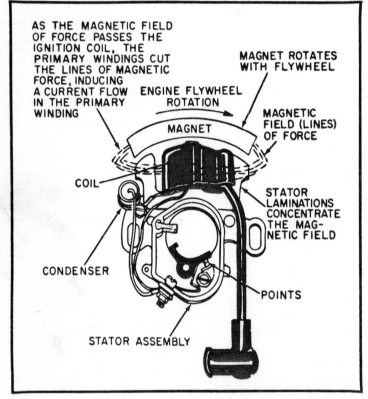

AS THE MAGNETIC FIELD OF FORCE PASSES THE IGNITION COIL, THE PRIMARY WINDINGS CUT THE LINES OF MAGNETIC FORCE, INDUCING A CURRENT FLOW IN THE PRIMARY WINDING

MAGNET ROTATES WITH FLYWHEEL

ENGINE FLYWHEEL ROTATION

MAGNET

MAGNETIC FIELD (LINES) OF FORCE

COIL

STATOR LAMINATIONS CONCENTRATE THE MAGNETIC FIELD

CONDENSER

POINTS

STATOR ASSEMBLY

Fig. 4-8 Magneto parts arrangement. The cam which opens the points is not shown.

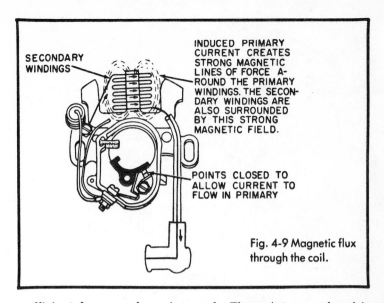

SECONDARY WINDINGS

INDUCED PRIMARY CURRENT CREATES STRONG MAGNETIC LINES OF FORCE A-ROUND THE PRIMARY WINDINGS. THE SECON-DARY WINDINGS ARE ALSO SURROUNDED BY THIS STRONG MAGNETIC FIELD.

POINTS CLOSED TO ALLOW CURRENT TO FLOW IN PRIMARY

Fig. 4-9 Magnetic flux through the coil.

sufficient for normal service work. The points are closed in Fig. 4-10 and the primary circuit is completed to ground. Since the primary conductor is in a moving magnetic field (the flywheel is turning), current is generated in the primary winding. Primary current generates a magnetic field of its own which envelops the secondary windings. The points open and the primary field collapses across the secondary. Again we have the familiar phenomenon of a conductor in a moving magnetic field. The disproportionate ratio between the number of turns in the two windings causes a high-voltage, low-amperage current to be induced in secondary. Flywheel magnetos can deliver between 18,000 and 21,000 volts to the spark plug.

Magneto Troubleshooting and Repair

Remove the spark output by holding the lead ⅛ to ¼ in. from the block; it should be thick and blue at cranking speed. You should have the tools shown in Fig. 4-11, but most mechanics do not bother to use them. Hold the wheel with one hand and drive the clutch off using a soft wood block against one of the ears.

The flywheel is secured to the shaft by a taper and a key. As mentioned in Chapter 2, the safest way to remove the wheel is with a puller designed to engage holes in the hub. If you want to use a knocker, do so at your own risk. Screw the knocker

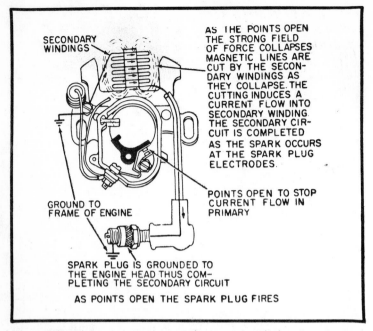

AS THE POINTS OPEN THE STRONG FIELD OF FORCE COLLAPSES MAGNETIC LINES ARE CUT BY THE SECONDARY WINDINGS AS THEY COLLAPSE. THE CUTTING INDUCES A CURRENT FLOW INTO SECONDARY WINDING. THE SECONDARY CIRCUIT IS COMPLETED AS THE SPARK OCCURS AT THE SPARK PLUG ELECTRODES.

SECONDARY WINDINGS

GROUND TO FRAME OF ENGINE

POINTS OPEN TO STOP CURRENT FLOW IN PRIMARY

SPARK PLUG IS GROUNDED TO THE ENGINE HEAD THUS COMPLETING THE SECONDARY CIRCUIT

AS POINTS OPEN THE SPARK PLUG FIRES

Fig. 4-10 Points open, secondary windings energized (drawings courtesy Tecumseh Products Co.)

down finger tight and hit it with an 8 oz hammer. Should the wheel be reluctant, pry on the backside with a screwdriver and hit it harder. Strike dead-on; a glancing blow can break the crankshaft stub. A wheel which does not respond to outward pressure from the screwdriver, or impact from the knocker

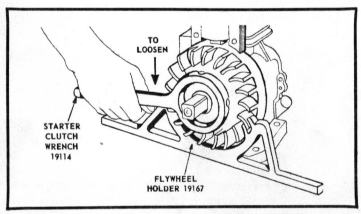

TO LOOSEN

STARTER CLUTCH WRENCH 19114

FLYWHEEL HOLDER 19167

Fig. 4-11 Removing flywheel nut (Courtesy Briggs & Stratton)

has probably suffered keyway damage and should be lifted with a puller. Some mechanics use heat to expand the hub, but this is dangerous unless you know the location of the coil.

Inspect the flywheel hub for cracks. Figure 4-12A illustrates damage caused by sudden engine stoppage. This sort of damage is most frequently encountered on direct-drive rotary lawnmowers, but has been known to occur among the other types. Drawing B shows damage caused by a loose flywheel. Both of these conditions can shear the key; the direction of shear depends upon whether blade inertia, or engine torque, was the culprit.

Assuming that the flywheel and crankshaft keyways are square and true, continue your inspection with the rim of the wheel, looking for broken fins and worn starter engagement teeth. A wheel with a broken impellor fin should be discarded. It can be salvaged for light service, however, if an opposed fin is snapped off to maintain balance. (Of course, expect pumping efficiency to be down.) Some starter ring gears are replaceable; others (like the one illustrated in Fig. 4-12) are cast integrally with the wheel.

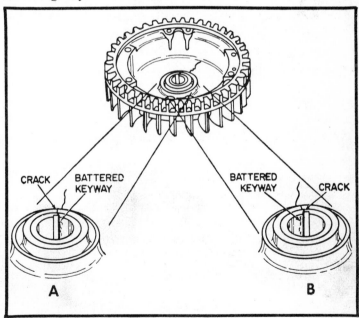

Fig. 4-12 Flywheel damage. A: inertial damage; B: insufficient torque on the nut (Courtesy Tecumseh Products Co.)

Contact Points

Remove the dust cover and inspect the points, turning the shaft until the points open fully. A gap specification of 0.020 in. has been standard on light utility engine magnetos since 1948, although some Clinton 2-cycles seem to run better with an 0.018 in. gap.

Pay particular attention to the condition of the contact surfaces. A mottled appearance is normal, but pitting, or the transfer of metal from one to the other, are signs of impending failure. The contacts should be slate gray—lighter colors, indicating overheating and carbon deposits, mean that the upper crankshaft seal is leaking excessively. Overheating can also turn the tip of the movable contact arm blue. If some tungsten remains, the contacts can be restored by filing. Remove the points from the engine and support each component in a vise; tungsten is a very hard metal and requires considerable filing. Since it is practically impossible to file both surfaces dead parallel, round them slightly, the highest point at the center. Do not use a stone—particles will imbed themselves in the contacts and cause local overheating. Finish the work with a fine single-cut file. Points reconditioned in this manner will not last as long as new ones.

Briggs supplies a nylon tool for depressing the spring at the primary connection (Fig. 4-13C). The spring can be turned over the wire by hand.

After installation, check for full contact between the points. The ideal figure is 100%, but 70% overlap is adequate (Fig. 4-14). Bend the fixed point bracket and set with a feeler gage (or with the tab on the dust cover of some magnetos). A wire gage is convenient since it does not have to be held parallel to the point surface. Bracket your measurement by using two blades on a machinist's gage. If the specification is 0.020 in., a 0.019 in. blade should slip through without drag, and a 0.021 in. blade should meet some resistance. Recheck the points after final tightening. Burnish the contacts with a card to remove fingerprints or oil that might have been on the gage. Failure to do this could cause the engine to refuse to start.

A weak return spring can cause high-speed misfiring. If you suspect this is the case, replace the points with factory originals. Do not use after market ignition components in lawnmower magnetos.

The movable arm is "hot" and must not be grounded. Certain designs had a long spring which wound around the

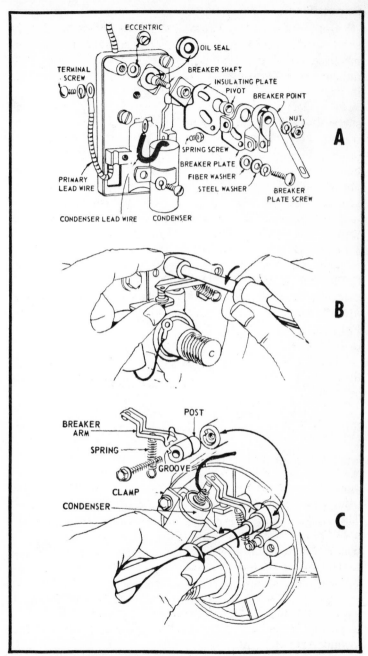

Fig. 4-13 Two Briggs & Stratton point assemblies. The set shown at A and B is used on horizontal-shaft engines (Courtesy Briggs & Stratton)

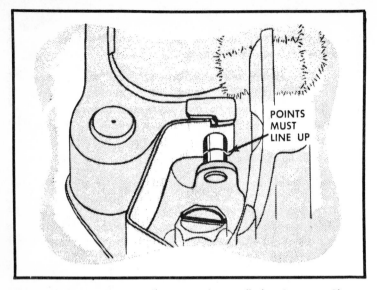

POINTS
MUST
LINE UP

Fig. 4-14 The points must line up and meet flatly (Courtesy Clinton Engines Corp.)

inside of the point chamber. Unless the mechanic is careful, the spring will pivot and touch metal as the condenser and coil connections are made. Primary wiring to the points may also tangle with the flywheel if not tucked clear of rotating and oscillating parts.

Most point arms are in direct contact with the cam; Briggs and a few Clinton magnetos employ a pushrod between the cam and movable arm. The pushrod should be inspected for wear on the sides and at the cam end.

Ignition Cams

Cams rarely give trouble as long as they receive some lubrication and the dust covers remain intact. Most cams slip over the crankshaft stub and are installed as shown in Fig. 4-15. Briggs & Stratton machines have a flat on the crankshaft to allow the points to close. Severe wear, or scoring, means crankshaft replacement. However, I have never seen a Briggs crank fail in this manner. Usually they can be burnished and reused with a new pushrod.

Condensers

Condensers, known in the electronics trades as capacitors, are devices that trap electrons. Wired across the points, the

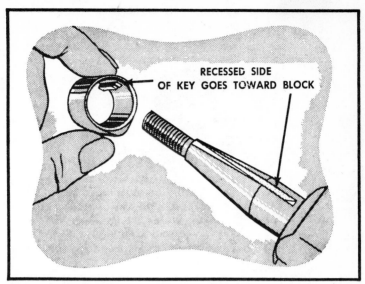

Fig. 4-15 Replaceable cam. Note the position of the recess. (Courtesy Clinton Engines Corp.)

condenser momentarily stores electrons which would arc and burn the contacts. Condenser construction is shown in Fig. 4-16. The cam is grounded to the stator plate, and isolated from the inner foil by plastic or waxed paper insulation. Electrons are attracted to the foil by the proximity of ground. The foil area and the thickness of the insulation (or dielectic) determine electrical capacity. Small engine condensers are rated at between 0.19 and 0.29 μF. (microfarads). Undercapacitance (for a given condenser—coil combination)

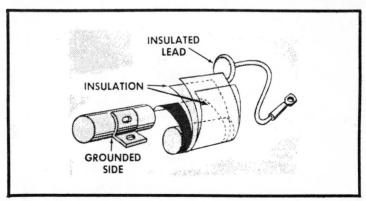

Fig. 4-16 Condenser construction (Courtesy Clinton Engines Corp.)

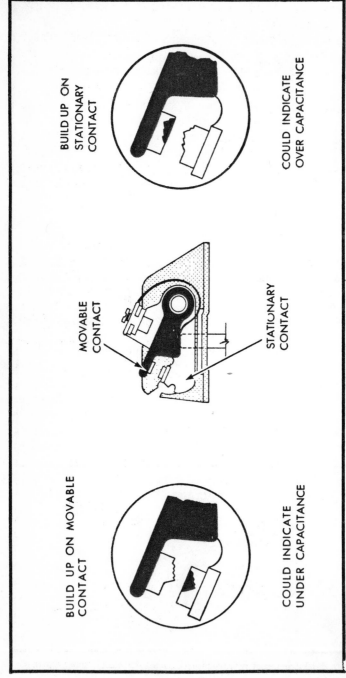

BUILD UP ON
MOVABLE
CONTACT

COULD INDICATE
UNDER CAPACITANCE

MOVABLE
CONTACT

STATIONARY
CONTACT

BUILD UP ON
STATIONARY
CONTACT

COULD INDICATE
OVER CAPACITANCE

Fig. 4-17 The condenser must be matched to the resto of the system (Courtesy Kohler of Kohler)

can cause point metal to migrate to the movable contact (Fig. 4-17). Overcapacitance does the reverse.

The symptoms of condenser failure are burned or cratered points, low voltage output, and excessive sparking at the contacts. You can sometimes see evidence of sparking by discoloration of the stator plate under the contacts.

While condensers can be checked with a high-range ohmmeter, and their capacitance determined by any of the several magneto testers on the market, the best approach is to change the condenser whenever the points are replaced. Briggs makes this unavoidable on many models by making the condenser integral with the grounded point contact (Fig. 4-13B).

Coils

Coil failure can result in no spark, weak spark, or intermittent spark. In the course of the troubleshooting procedure, coil diagnosis is left until last. If new points and a new condenser do not solve the problem, the technician checks the primary wiring—remember, most lawnmowers have a shutoff switch outside of the magneto proper. Postponing coil

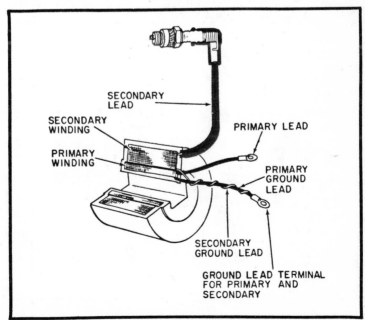

Fig. 4-18 Magneto coil construction (Courtesy Tecumseh Products Co.)

analysis is justified by the rarity of coil malfunctions and the difficulty of performing accurate coil tests. Magneto analyzers do not always tell the truth about coil performance. The surest test is to substitute a coil known to be good that has the same part number.

Most American coils—Bendix, Phelon (Repco), Scintilla, Wico—are secured to their cores, or pole shoes, by spring taps (see Fig. 4-21). Once the tap is pried free with a screwdriver, the coil should slip from the core. If it doesn't, place the assembly in a vise (jaws supporting the coil), and gently drive the core down and out. This method *must* be used with late-production coils which are bonded to their cores. **Do not apply force to the coil itself**.

Before assembling a new coil over the core, note the position of the ground and primary leads. Push the coil over the laminations *with hand pressure only*. Insert the locking tab or use epoxy. Do not solder the high-voltage (spark plug) lead to the coil clip.

the core laminations may be damaged if they have been allowed to come into contact with the flywheel rim. The radial distance between the core and the flywheel is known as the flywheel–coil gap, or air gap. Since magnetic force decreases

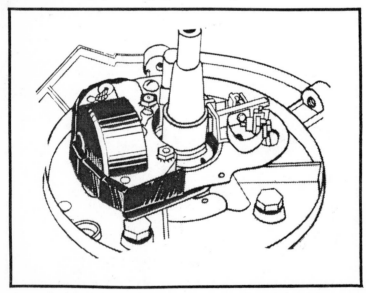

Fig. 4-19 Checking the air gap on magnetos with internal coils (Courtesy Clinton Engines Corp.)

faster than an increase in distance from the source would imply, the air gap should be as close as possible consistent with running clearance. Worn main bearings, a bent crankshaft, or improper assembly can change the air gap. Magnetos with external coils, such as the one illustrated in Fig. 4-20, can be set with an ordinary business card or, more professionally, with a nonmagnetic feeler gage that is normally used for clutch adjustments on Delco air-conditioning compressors (available from some automotive supply houses).

Magnetos with internal coils are usually not adjustable, unless one takes the trouble to grind the pole shoes or to drill the mounts oversize. The latter method is slow, and care must be taken not to spread the laminations. This technique is used on the few surviving engines with magnetic pole shoes. Check the clearance by wrapping successive layers of vinyl tape over the core as shown in Fig. 4-19. When applied flat, with just

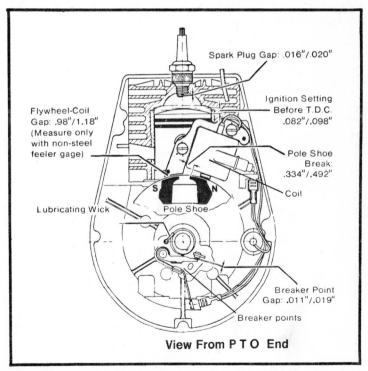

Spark Plug Gap: .016"/.020"

Ignition Setting Before T.D.C. .082"/.098"

Flywheel-Coil Gap: .98"/1.18" (Measure only with non-steel feeler gage)

Pole Shoe Break: .334"/.492"

Coil

Lubricating Wick

Pole Shoe

Breaker Point Gap: .011"/.019"

Breaker points

View From P T O End

Fig. 4-20 A Robert Bosch magneto, showing various adjustments (Courtesy Rockwell Mfg. Co.)

enough tension to prevent wrinkles, vinyl tape (when measured with a micrometer) affords a thickness of 0.008 to 0.009 in. per layer. If the wheel drags when torqued down on one layer, the gap is too narrow. Three or more layers before contact means the gap is too wide for best spark.

The pole shoe break (also known as the breakaway gap, edge gap, or E-gap) refers to the position of the core relative to the flywheel magnet at the moment the points open. Bosch and other foreign magnetos provide an adjustment for this gap; American magnetos are set at the factory. A small variation from specified tolerances will result in a weak spark or no spark at all. Check the flywheel key and keyway, then replace the cam. As a last resort, experiment with point settings to obtain maximum spark at cranking speed. In reference to Fig. 4-19, widening the point gap narrows the pole shoe break. Depending upon the diameter of the flywheel, a point gap change of 0.001 in. can move the pole shoe break by $1/32$ inch.

Ignition Timing

Most lawnmower engines have a preset static advance. This category includes Briggs & Stratton, Clinton, and Pincor utility engines. Tecumseh engines can be timed in the field. The stator plate is rotated until the punch mark aligns with the raised rib on the block mounting fixture. A variation of this is shown in Fig. 4-21. Tecumseh 2-strokes may have paired

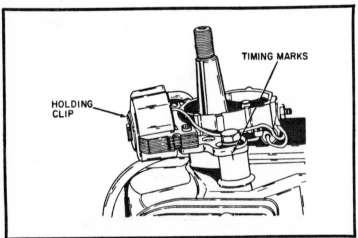

Fig. 4-21 Typical Lauson timing marks (Courtesy Tecumseh Products Co.)

timing marks, or may use the open ends of the mounting fixture as a reference point. These marks are a service convenience and are not intended to be absolute guides. The only way to accurately time an engine is to set the points at the specified gap, and rotate the crankshaft to a specified position before top dead center (TDC). The points should barely open. Most timing instructions are given in decimal fractions of an inch of piston travel before TDC. Piston movement may be determined by a machinist's rule inserted into the spark plug boss (Fig. 4-22), by a graduated scale which screws into the boss (illustrated in Fig. 2-10—available from small-engine distributors, as well as from Tecumseh); for absolute accuracy, a dial indicator is the way to go. Whether the cost of a dial indicator can be justified for lawnmower maintenance is questionable; many small-engine mechanics have never learned how to use one. However, there is some satisfaction in doing a job without the usual cost and time compromises. If you would like to invest in a dial indicator, your best bet would likely be one of the general-purpose types—useful to detect crankshaft, blade, and clutch runout, as well as to establish piston movement.

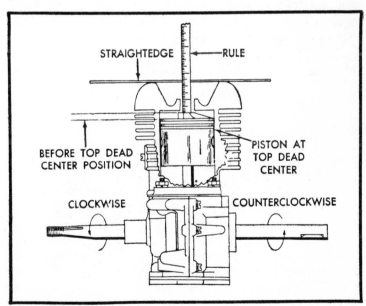

Fig. 4-22 Using a rule and straightedge to time a Power Products engine (Courtesy Tecumseh Products Co.)

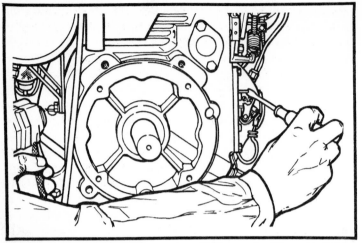

Fig. 4-23 Using a strobe light to time a Kohler engine (Courtesy Kohler of Kohler)

A few manufacturers give timing specifications in degrees of crankshaft rotation before TDC. The crude method is to count the impellor vanes on the flywheel and divide into 360. For example, if there are 20 vanes, the spacing between centers is 18°. Half the distance is 9°. A better way is to use a degree wheel available from specialty automotive suppliers.

Some of the more ambitious engine designs use a separate, self-contained magneto driven by the camshaft. Others employ flywheel magnets but have camshaft-operated points. In either case, the points are driven at half engine speed and do not produce the twice-as-many sparks necessary for 4-cycles. The cam is an ideal location for the ignition-advance weights and return springs. A few foreign engines have an advance mechanism tucked under the flywheel, but space is limited and dust accumulations are a problem in lawnmower service.

In general, service and inspection procedures parallel those used for flywheel magnetos. Figure 4-23 shows a strobe light used to time a Kohler single-cylinder engine. Two marks—prefix T or S—are stamped on the inner side of the flywheel. With the engine turning at 1200 rpm, the S mark should align with the reference line. Minor adjustments are made by changing the point gap .002 in. either side of the specified 0.020 inch. Major adjustments require that the point chamber cover be removed and the point assembly bracket rotated on its mounting screws. The flywheel should be positioned at S. Static timing is completed when the points just break. Tighten the bracket screws and check the timing with a light. It is not necessary to invest in an expensive *dc* strobe. A neon timing light will do as well, so long as test are not made in direct sunlight. As a timing aid, lighten the S mark with chalk.

Kohler engines have a provision for spark advance or for neutralization of cranking "kickback." With the exception of the K91, all single-cylinder models feature an automatic compression release; at low speeds the cam holds the exhaust valve open. With this system, full running advance can be used during starting. The K91 engine employs spring-loaded weights between the camshaft and the point drive mechanism. With increased speed, the weight (on a lever) is thrown outwards, advancing the point break relative to piston travel. Since flame-front velocities in the chamber rarely exceed 300 ft/sec, ignition has to be initiated ever earlier as piston speed increases. The operation of the automatic advance can be verified with a strobe light.

Large Briggs & Stratton engines have three distinct magneto models (although some components are interchangeable). Two types, neither of which was given a marketing name, have external coils and flywheel magnets. The third, known as the Magna-Matic, has an under-the-sheel coil and a rotor.

Point gap is 0.02 in. for all models. Early-production external coil magnetos have a pole shoe break timing mark as drawn in Fig. 4-24. The pole shoe can be moved laterally on its

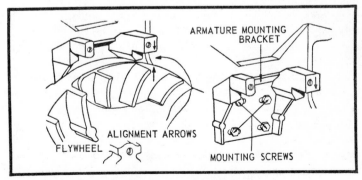

Fig. 4-24 Early production pole shoe break marks. Align the arrows at the moment of point break (Courtesy Briggs & Stratton)

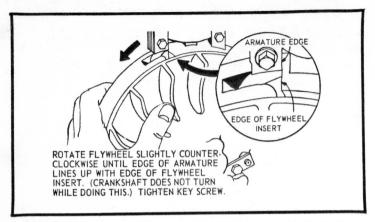

ROTATE FLYWHEEL SLIGHTLY COUNTER-CLOCKWISE UNTIL EDGE OF ARMATURE LINES UP WITH EDGE OF FLYWHEEL INSERT. (CRANKSHAFT DOES NOT TURN WHILE DOING THIS.) TIGHTEN KEY SCREW.

ARMATURE EDGE

EDGE OF FLYWHEEL INSERT

Fig. 4-25 Current production pole shoe break (Courtesy Briggs & Stratton)

mounting screws to align the arrows at the moment of point break. The air gap between the pole shoe and the outer rim of the flywheel should be 0.01 inch. Later model external coil magnetos can be identified by a wedge-shaped flywheel key which is secured to the wheel with a hexhead screw. Pole shoe break is set by turning the flywheel (key installed) until the points just open. Remove the key and turn the wheel on the crankshaft taper until the armature edge aligns with the flywheel insert. Be careful not to turn the shaft during this operation, since shaft rotation will open or close the points beyond initial break. The air gap between the pole shoe and the flywheel is 0.022 to 0.026 inch.

The Magna-Matic rotor is secured to the crankshaft by a pinch bolt or a setscrew. There should be 0.025 in. of clearance between the back of the rotor and the main bearing support. The pole shoe air gap is fixed. The crank or the main bearings should be replaced if you can insert a 0.004 in. gage blade at any point between the rotor and pole shoe. (See Fig. 4-26.) Timing varies between engines; rotors are coded by model number. Set the points at 0.02 in., and rotate the crank in the normal direction to point break. Align the model number reference line with the arrow (Fig. 4-27).

MAGNETO ASSEMBLY

Recheck all clearances, lubricate the cam with a very thin film of high temperature grease such as Delco cam lubricant or Alvania No. 3 (a German Shell product available from

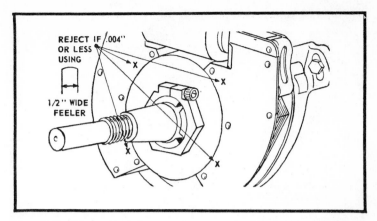

Fig. 4-26 "Magna-Matic" rotor clearance checks (Courtesy Briggs & Stratton)

Sachs dealers). If a wick is provided, position it so that it lightly wipes the cam with three or four drops of motor lube. Install woodruff keys as demonstrated in Fig. 4-28. The edge of the key should be parallel to the crankshaft axis. Set the flywheel over the taper and, with the spark plug out, spin the engine by hand, being careful not to gash your fingers on the governor air vane. The magneto should deliver a thick, blue spark. If all adjustments have been made properly, you will be able to hear a snap as the spark discharges to ground. The absence of a spark usually means that the points have been contaminated by fingerprints or oxidation. New points are not

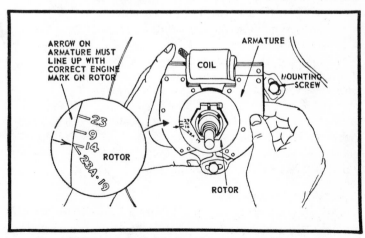

Fig. 4-27 "Magna-Matic" timing (Courtesy Briggs & Stratton).

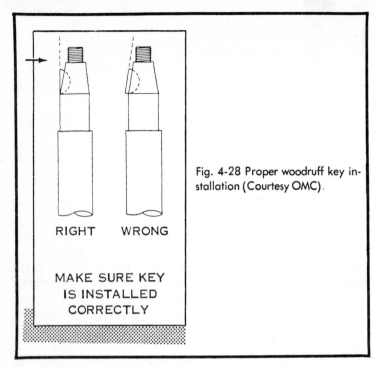

Fig. 4-28 Proper woodruff key installation (Courtesy OMC).

RIGHT WRONG

MAKE SURE KEY
IS INSTALLED
CORRECTLY

immune to deterioration in storage. Burnish the contacts and try again.

FLYWHEEL TORQUING SPECIFICATIONS
BRIGGS & STRATTON ALUMINUM BLOCK

	TORQUE
6B, 60000, 8B, 80000, 82000, 92000	57 ft-lb
1000000, 130000	60 ft-lb
140000, 170000, 190000	67 ft-lb

Cast Iron Block

5, 6, N, 8	57 ft-lb
9	60 ft-lb
14	67 ft-lb
19, 190000, 200000	114 ft-lb
23, 230000, 243400, 300000, 320000	144 ft-lb

CLINTON
Gem, Clintalloy, Long Life   flat washer—200–250 in.-lb
                split; compression; lockwasher—350–400 in.-lb
                    (⅞ in. crankshaft 350 in.-lb maximum)

KOHLER SINGLE CYLINDER
K91	40 – 50 ft-lb
K141, K161, K181	50 – 60 ft-lb
K241, K301, K321, K341	60 – 70 ft-lb

OMC C SERIES, D SERIES   330 – 370 in.-lb

TECUMSEH
Four-Cycle, 1¾ hp – 5 hp (inclusive)   30 – 33 ft-lb
Two-Cycle, right- or left-hand thread   18 – 25 ft-lb, 10 ft-lb
                minimum after run and cooldown.

BATTERY IGNITION SYSTEMS

Engines used on garden tractors and riding mowers sometimes incorporate a battery-and-coil ignition. This system is similar in operation to the magneto, except that primary current is provided by a storage battery. At the moment of point break, high voltage current is induced in the secondary windings in response to the collapse of the magnetic fields around the primary. Service operations differ little from those required for magneto systems. Figure 4-29 is a pictorial schematic of a typical battery – coil design.

SOLID-STATE IGNITION SYSTEMS

Battery-and-coil ignition is somewhat cumbersome and falls short at high speeds. Normally, the system has enough reserve so the voltage drop is not critical; but the potential for high-speed misfire is inherent.

Magnetos are compact and perform best in sustained high-speed operation—but at lower engine speeds, the spark is weak, since output depends on the relative velocity of the magnetic field and the primary winding. At cranking and sudden overloads—exactly when we need a spark the most—magneto output drops, and the ignition points become especially troublesome. Magneto primary currents are higher than battery – coil ignitions used in automobiles. Point life, which in the last analysis is a function of the amount of current arcing, is quite short. Consequently, lawnmowers have to be serviced every few hundred hours. Apart from burning, points change their gap with wear against the cam and rubbing block. Gap changes translate as timing and pole-shoe break changes.

Small engine manufacturers have been slowly phasing out magneto and battery – coil systems in favor of solid-state systems. The technology has been proven in outboard,

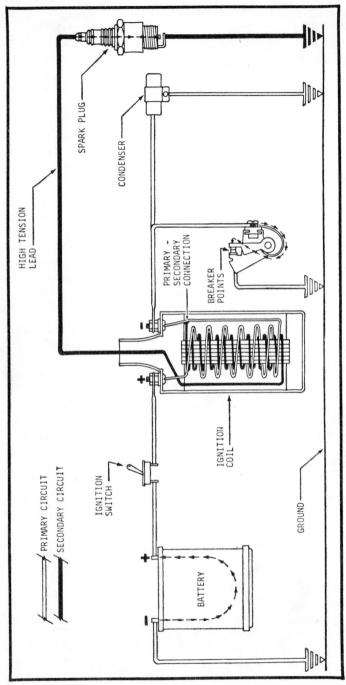

Fig. 4-29 Battery and coil ignition (Courtesy Kohler of Kohler)

snowmobile, and motorcycle applications. One could hazard the guess that if it were not for the increased cost of solid-state systems, the magneto would long since have gone the way of the starter cord.

The Kohler capacitive discharge ignition (CDI) is typical of the new breed. It is shown in schematic form in Fig. 4-29. 30 Current generated in coils (mounted on the stator plate) alternates because of shifting magnetic polarity. It changes direction several hundred times a minute. Half of the output charges a large capacitor (the technically correct name for a condenser) shown at the right of the schematic. The diodes are electrical traffic cops; these semiconductors permit current to move in only one direction (actually, against that indicated by its symbol). The low resistance entry point is shown in the schematic as the cross-barred tip of the arrow.

A charge of electrons moving from the top of the ac winding will be blocked by diode 2. It returns to the winding, completing the circuit, through diode 1. A charge moving in the opposite direction will be blocked by diodes 1 and 3. It must travel through the ignition coil primary where it charges the capacitor. The capacitor continues to collect electrons on each half-alternation. Electrons cannot find their way to the other plate to equalize the charge because diode 2 stands in the way. Were it not for this diode, the capacitor would discharge through the ac winding.

The charge is released through the action of the sensing (or trigger) coil and the SCR (silicon controlled rectifier). The sensing coil is a small coil mounted on the stator with the ac windings. It develops a small voltage when swept by the magnetic field. This voltage goes to the SCR. The SCR is a switch without moving parts. It conducts when signaled by sensing-coil voltage. The capacitor discharges, flooding the primary winding of the ignition coil, inducing tremendous voltages in the secondary winding.

Other features of this circuit are the resistor, which dampens stray currents, and diode 3, which prevents the SCR from being triggered by capacitor voltage. The switch bypasses the ignition coil and capacitor.

In addition to high voltage—some manufacturers say hazardously high—CDI systems have a very low rise time—on the order of 2 microseconds (Fig. 4-30). The spark plug may leak like the proverbial sieve, but the voltage rise is so rapid that the plug does not have time to short to ground. A good CDI

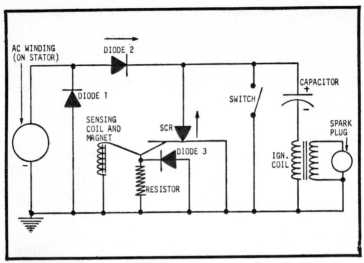

Fig. 4-30 CDI for small engines (Courtesy Kohler of Kohler)

system will fire a plug that has been dipped in oil and had its side electrode removed.

For all their advantages, and although they most certainly represent the wave of the future, CDI systems are not perfect. Their circuits are integrated into sealed modules; in the event of failure, the affected module must be replaced as a unit, and

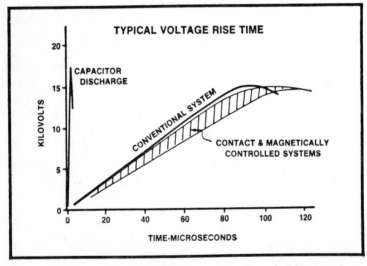

Fig. 4-31 Comparison between conventional and CDI rise time (Courtesy Champion Spark Plug Co.)

parts costs are relatively high. Overall plug life is extended, especially in 2-cycle applications, but electrode erosion has been a problem with these units in small engine applications. High-voltage cables tend to leak with age and must be renewed more often than normal for conventional systems. Radio interference has also been a problem. For this reason CDIs are banned in parts of Europe and Japan.

Spark Plugs

Spark plug components are nomenclatured in Fig. 4-32. The center electrode has a terminal that connects to the high-voltage lead from the coil secondary. The side (or ground) electrode is made of special alloy to resist "washout" by combustion gases. All modern electrical system components have a negative ground, except for the spark plug—it is positively grounded. That is, electrons move from the hot center terminal to the relatively cool side terminal. Electron emission is in the direction of temperature dissipation.

The insulator is made of ceramic. It has several million ohms of resistance and can withstand severe thermal shock. On a winter day the terminal end may be chilled below freezing while the firing tip is at 1000°F. However, the insulator is mechanically brittle, as anyone who has slipped a wrench against one can testify. The shell is usually made by the cold extrusion process. It supports the insulator and makes a

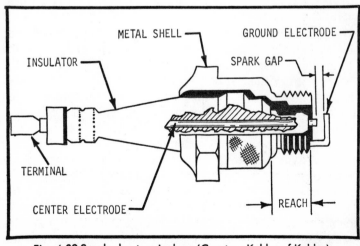

Fig. 4-32 Spark plug terminology (Courtesy Kohler of Kohler)

heatsink (heat dissipator) between the firing tip and cylinder head. Some shells have a tapered seat, as shown in the drawing, and require no gasket. The standard thread size is 14 mm. Reach varies between $\frac{3}{8}$ and $\frac{3}{4}$ in. to accommodate various engines. Nearly all lawnmower engines use those plugs with the shortest reach. A longer reach substitute may extend the firing tip into the piston path and will disturb the symmetry of the chamber. Unless otherwise specified, the gap should be 0.025 inch. Most mower plugs accept a $^{13}/_{16}$ in. socket, although the newer bantam plugs require a $\frac{5}{8}$ or $\frac{3}{4}$ in. wrench.

Torque values are subject to modification by the various engine builders; the following chart is provided as a guide:

| | TORQUE VALUES | |
PLUG THREAD DIAMETER	ALUMINUM	CAST IRON
14 mm tapered	10 − 15 ft-lb	10 − 15 ft-lb
14 mm standard	18 − 22 ft-lb	26 − 30 ft-lb

When installing a plug, clean the seat, dab a bit of antiseize compound on the threads, and torque to the lowest value shown. If there is evidence of leakage, allow the head to cool and torque to the highest value.

In addition to thread diameter and reach, spark plugs exhibit a variety of tip and electrode styles. Small design changes can reduce electrical interference and extend cleaning intervals. But the most important parameter is the plug's heat range. Ideally, the firing tip should be kept at approximately 1000°F. Below this temperature the plug tends to carbon foul; higher temperatures cause preignition. The location of the plug, the geometry of the chamber, the temperature of the head, compression ratio, rpm, and intended service—all influence the choice of the heat range. An engine which develops high combustion temperatures requires a cold plug. Conversely, one which tends to run cool should have a hot plug.

The heat range is a function of the shape of the insulator nose. As you can see from Fig. 4-33, a cold plug provides a relatively short heat path from firing chamber to head. A hot plug forces heat to travel further. Consequently, the firing tip is hotter for the same temperature input. The length of the heat path is not a solitary factor—some insulators have a greater area exposed to combustion than do others. The per-unit thermal load is greater, making the tip correspondingly hotter.

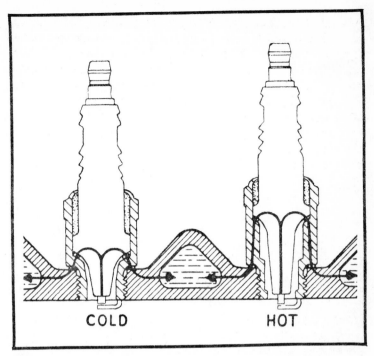

COLD   HOT

Fig. 4-33 Although other factors are involved, the distance heat travels from the nose to the shell is the primary determinant of the plug's heat range (Courtesy Champion Spark Plug Co.)

Race car mechanics juggle heat ranges, in a search for the ideal plug to match the course, car, and driver. There is no similar justification to deviate from manufacturer's specifications for lawnmowers; in fact, there's danger in the practice—a hot plug can destroy the engine in a few minutes under load. Without the plug suggested by the manufacturer, there is no cross check on ignition, carburetion, and other variables. The Champion plug chart (Fig. 4-34) lists recommendations for all popular mowers. The J-17LM is standard for approximately 90% of mowers currently in production.

Reading Spark Plugs

The only way a mechanic can tell what's been going on in the combustion chamber is to examine the spark plug firing tip. The tip responds to fuel chemistry, temperature, and to the presence of oil. A mechanic who is well versed in the art of "reading" plugs can describe the history of the engine and

ACME...... J-8 or J-17LM
AIR CAP 14mm Heads..........J-17LM
ALDENS (gap .030").....J-8 or J-17LM
ALLIS-CHALMERS ...J-8 or J-17LM
AMERICAN....J-7 or J-17LM
ARTISAN (Gambel)......J-8 or J-17LM
ATCO....................D-16 or UD-16
ATLAS......J-8, CJ-8 or J-17LM
B-M................J-8 or J-17LM
BEAZLEY 14mm Heads...J-8 or J-17LM
  18mm Heads............D-16 or UD-16
BIG SNAPPER........D-16 or UD-16
BLAIR 14mm Heads.........J-17LM
BOB-A-LAWN 14mm Heads...J-17LM
BOLENS Orbit-Air & Suburban......J-8
BUCH 14mm Heads........J-17LM
BUNTON (All gaps .030")
  Model B..........J-8J or J-17LM
  Little Mo, Red Wing, Dilly 2,
  Duzmore 16".............J-12J
  Dilly 4, Duzmore 18", C-18, C-21,
  Lawn Lark........J-8 or J-17LM
CADILLAC Clipper...........UY-6
CASEY MOW 14mm Heads.....J-17LM
CHALLENGE...................A-25
CHIEF using Power Pak..J-11 or J-17LM
  Others (gap .037").............UJ-12
CHOREMASTER 14mm Heads....J-17LM
CLEAN-CUT..........J-8 or J-17LM
CLEMSON 14mm Heads........J-17LM
COLDWELL H30, G35, L25, M40
  using Flint & Walling, Grey......A-25
  H30, G35, R25 using:
  Fuller & Johnson..............W-18
  Coldwell Engine..............W-14
  B & S, Clinton.....J-8 or J-17LM
CON-SOL 14mm Heads.........J-17LM
CONTOUR 14mm Heads........J-17LM
COOPER 14mm Heads.........J-17LM
CRAFTSMAN 14mm Heads.....J-17LM
CUNNINGHAM 14mm Heads.....J-17LM
DAVIS................J-8 or J-17LM
DUNLAP (Sears) using B & S........J-8
  Power Products........J-11J or HO-8A
DUO-THERM 14mm Heads....J-17LM
EASY WAY 14mm Heads......J-17LM
ECLIPSE..........J-8 or J-17LM
ESKA 14mm Heads.........J-17LM
EVERSHARP 14mm Heads....J-17LM
EXCELLO using Power Products...J-11J
  All others.............J-8 or J-17LM
EXPERT.............J-11 or HO-8A
E Z 14mm Heads...........J-17LM
F & N 14mm Heads.........J-17LM
FAIRBANKS MORSE 14mm Hds..J-17LM
FALLS ROTO CLIPPER using:
  Power Products Eng.........J-11J
  All others.......J-8 or J-17LM

FEDWAY B & S Eng....J-8 or J-17LM
  Iron Horse Eng................J-11J
FOLEY 14mm Heads............J-17LM
FOX 14mm Heads............J-17LM
GAMBLE-ALDENS
  (gap .030")........J-8 or J-17LM
GARDEN KING................D-14
GARDEN MART (Wards)
GOODALL ROTARY using:
  McCulloch or Lauson Eng J-8 or J-17LM
  Others (gap .037")..... • ...UJ-12
GRAVELY Clean-Cut Model 24......J-8
HOME LAWN (See "Toro")
HOMELITE "Yard Trac" (gap .030")..J-8
  M26 (gap .025")..............J-6J
HOMKO 14mm Heads.......J-17LM
HUFFMAN
  Mowers, Tillers, Edgers...CJ-8 or J-8
HUFFY.........J-8 or J-17LM
HURRICANE 14mm Heads.......J-17LM
IDEAL ⅞" Heads.............W-10
  18mm Heads.........D-16 or UD-16
  14mm Heads.........J-17LM
INTERNATIONAL HARVESTER
  Riding Mower Cadet 60.......J-8
JACKSON 14mm Heads.......J-17LM
JACOBSEN ½" Heads..........A-25
  ⅞" Heads................W-14
  4-Acre Heavy Duty
   (gap .035").......UD-16 or D-21
  All other 18mm Hds (gap .030")...D-16
  Reel Type 14mm Heads
   Briggs & Stratton Eng...CJ-8 or J-8
   All others.............UJ-12*
  Rotary Type, 14mm Heads
   B&S, Clinton, Lauson....CJ-8 or J-8
   Jacobsen................J-8J
  Mow-Mobile, Park 30,
  Turf Commander, Turfgroomer..D-16J
  Turf King, Javelin
   w/B&S Engs........CJ-8 or J-8*
   w/Kohler Engs.............H-10
  Putting Green, Greens Mowers..UJ-12*
  Model F, Greens King,
  Commercial 60.............H-10
  Model G...............UD-16
  *Models w/J321 Eng. using
   "Shorty" plug use..........TJ-8J
  D10, E10, F10, G10 w/Ford Eng..H-10
  Also See Garden Tractors
JARI B & S Engines...........CJ-8
  Clinton 500.................J-12J
  Kohler & Tecumseh (gap .030")...J-8
JOHNSTON 14mm Heads.......J-17LM
JOLLYHOE 14mm Heads......J-17LM
KING O'LAWN......J-8 or J-17LM
KUT-KWICK
  B & S Engines.............J-8
  Wisconsin Engines...........D-16J
LAWNBOY B & S, Tecumseh..CJ-8 or J-8
  All other models............CJ-14
LAWNCRAFT................J-8
LAWN CRUISER (Canada)......UJ-12
LAWN-CYCLE 14mm Heads......J-17LM

LAWN MASTER 14mm Heads...J-17LM
LAZY BOY 14mm Heads.......J-17LM
LIMA Clinton Eng.............H-10
  Lauson Eng........J-8 or J-17LM
LOCKE 18mm Heads......D-16 or UD-16
  14mm Heads.......J-8 or J-17LM
MARAUDER using B & S Eng.......J-8
  Iron Horse Engine............UJ-12
MASSEY-FERGUSON All models...J-8
MC CULLOCH 14mm Heads...J-17LM
MEADE 14mm Heads..........J-17LM
MIDLAND 14mm 2-Cycle......CJ-14
  4-Cycle..J-8 or J-17LM
  18mm (gap .030")...D-16 or UD-16
MILWAUKEE................D-14
MILBRADT 14mm Heads..J-8 or J-17LM
  18mm Heads.........D-16 or UD-16
MONARK Special, Economatic
  Majestic Rotomatic..J-11J or J-17LM
MOTO MOWER
  14mm Heads........J-8 or J-17LM
  18mm Heads.......D-16 or UD-16
MOW-A-MAT 14mm Heads.....J-17LM
MOWAMATIC using:
  B & S 5S, 6S....J-8 or J-17LM
  Clinton VS200..............H-11
MOW-CYCLE 14mm Heads....J-17LM
MOW-MASTER using:
  Propulsion (gap .035").........J-11
  Power Pak (gap .035")..........J-8
  Clinton 2-cycle (gap .030")......H-11
MOW-MORE................UD-16
MOZ-ALL 14mm Heads......J-17LM
MUNCIE..........J-7 or J-17LM
MURRY (Schissel).....J-8 or J-17LM
NATIONAL................UD-16
NEW WAY.........UD-16 or D-23
PENNINGTON Using:
  B & S and Clinton Engs. (gap .028")..J-8
  Wisconsin Eng. (gap .028")......D-16
  Kohler..................H-10
  Onan...................H-8
PENNSYLVANIA Using Clinton Eng..H-10
  All other models......J-8 or CJ-8
PINCOR Using Pincor 4-cycle Eng. ...J-8
  Pincor 2-cycle Eng. (gap .035")....J-11J
  Briggs & Stratton...............J-8
  Clinton (gap .030")..........H-11J
PONY POWER........J-11 or J-17LM
POWER MATIC 14mm Heads....J-17LM
POYNTER...........J-8 or J-17LM
PRECISION-BUILT (Spiegel)......J-8
PRO Using:
  B&S or Tecumseh Engines......J-8
  3½ h.p. Clinton Engine.........H-10
  4½ h.p. Clinton Engine.........J-8

Fig. 4-34 Champion plugs for all popular mowers. Unless otherwise noted, the gap is 0.025 in.

RED BIRD............J-11J or J-17LM

RED CAP 14mm Heads.........J-17LM

REO 1949-1957.................UJ-12
  1958 on, See Com'l. Eng. Section

RIDE-A-MOWERS 14mm Hds...J-17LM

ROBERTON 14mm Heads.......J-17LM

ROBERTSON...........J-8 or J-17LM

ROOF using:
  Clinton, Kohler, Tecumseh Engs.....J-8
  Wisconsin Engine.......D-16 or UD-16

ROOT All models........J-8 or J-17LM

ROTH 14mm Heads............J-17LM

ROTOMATIC 14mm Heads......J-17LM

RPM Rotoflo...........J-8 or J-17LM
  Rototrim using: Clinton, B & S.....J-8
  Power Products.....J-11J or HO-8A
  Iron Horse..................UJ-12

RUGG 14mm Heads.............J-17LM

SARLO............CJ-8, J-8 or J-17LM

SAVAGE using B & S Engines .CJ-8 or J-8
  Clinton & Power Products 2-Cycle..J-11
  Clinton & Lauson Engines 4-Cycle...J-8

SCYTHETTE (gap .040")...........J-12J

SENSATION using B&S Engine......CJ-8
  Kohler Eng. (gap .030")........UJ-12
  Lauson Eng. (gap .030")..........J-8
  Cushman (gap .030").........UD-16
  Own Engine (gap .030").........H-10

Clinton (2-cycle) (gap .030").......H-11
Clinton 501 (gap .030")..........J-12J
Clinton 415....................J-8

SNAPPER TRIMMER....J-8 or J-17LM

SNAPPIN TURTLE 14mm Hds...J-17LM
  18mm Hds....D-16

SPIEGEL.............J-8 or J-17LM

SPINAWAY (gap .030").........D-16

STANDARD 14mm Heads..J-8 or J-17LM
  18mm Heads.................D-14

STEARNS 14mm Heads...J-8 or J-17LM
  18mm Heads..........D-16 or UD-16

STRUNK (gap .040")....UJ-12 or J-12J

SUNBEAM B & S 14mm Heads.......J-8
  18mm Heads.........D-16 or UD-16

SWISH-ERR 14mm Heads.......J-17LM

TORO ½" Heads.................A-25
  Tractor (Ford 6-cyl.)..............H-10
  Ford Shielded.............XEH-10
  Grass King.....................J-8
  Golfcourse Tractor.............W-18
  Other ⅞" Heads.............W-14
Parklawn, Professional and
  Mdl. MF 18mm Heads.........UD-16
  14mm Heads...............CJ-8
Park Special 14mm Heads.........J-8
  18mm Heads.............UD-16
Zipper, Homelawn, Starlawn, Power
  Greensmower, Sport Lawn, Greens-
  master, Super Pro, Big Red Rider..J-8
Bullet, General, Master, Parkmaster
  (mower tractors)................H-8
Whirlwind 18mm Hds. (gap .030").UD-16

Trojan (gap .030")...........D-16J

TRAMS 14mm Heads..........J-17LM

TRIMMER............J-8 or J-17LM

VOGT 14mm Heads.............J-17LM

VOLLRATH 14mm Heads........J-17LM

WARD LAKESIDE.....J-8 or J-17LM

WEGELE Scooter Mower...D-16 or UD-16

WHEEL HORSE...................J-8

WHIRL CUT 14mm Heads......J-17LM

WHIRLER-GLIDE........J-8 or J-17LM

WHIZ-MOW......................H-10

WIZARD (Western Auto) using:
  Propulsion VE901R (gap .035")....J-11
  Clinton VS700...................J-8
  Power Products (gap .030").......J-11J
  Clinton 2-cycle (gap .030").......H-11
  Power Pak (gap .035").............J-8

WORCESTER using B & S Engs......J-8
  Clinton & Power Products 2-Cycle..J-11
  Clinton & Lauson 4-Cycle..........J-8

WORTHINGTON.................UD-16

YARDMAN with Lauson Eng.......CJ-8
  B & S Engines......J-8 or CJ-8

YARDSTER............J-8 or J-17LM

YAZOO 14mm Heads...........J-17LM

YETTER ROTARY......J-8 or J-17LM

YORK 14mm Heads.............J-17LM

Fig. 4-34 (continued).

predict the next time an overhaul will be needed. These matters take skill and long experience. The average lawnmower mechanic or owner only needs to learn the subject in broad outline.

Figure 4-35 should be examined carefully. It is a photograph of a plug that has been overheated. The insulator is bone-white and the side electrode may have gunmetal-blue temper marks near the weld. Assuming the heat range is correct to begin with, this plug is saying that:

1) The ignition is out of time, or that
2) The carburetion is lean; i.e., there is too much oxygen in the mixture, either because of misadjustment, a clogged jet, or an air leak in the induction tract. A remote possibility is
3) lubrication failure.

The next photo shows overheating pushed past the threshold marked by preignition. This kind of plug damage is rarely seen in a lawnmower engine—the rod or piston usually fails before this happens. The photo is significant because it so

Fig. 4-35 Overheating—a danger sign (Courtesy Champion Spark Plug Co.)

dramatically illustrates the heat gradient in a spark plug. The hottest part—the center electrode—has completely disintegrated. The shell end of the center electrode remains relatively cool, but the contact tip has melted.

Figure 4-37 shows a plug fouled with carbon. Wet, oily deposits on the tip denote excessive oil in the chamber. Oil will raise the burning temperature (since oil heat generation is greater than that of gasoline) but the increase is not excessive in a lightly stressed engine. Oil fouling in a 4-cycle indicates worn rings, an overfilled sump or, possibly, a clogged crankcase vent. In a 2-cycle it means a too rich oil/fuel mixture, and can be aggravated by long idle periods. Dry, fluffy deposits which wipe off easily indicate a rich air/fuel mixture. The symptom is quite obvious on a 4-cycle, but can be masked as an excessive oil symptom on a 2-cycle. Check the carburetor adjustments and the choke for full opening. Unlike overheating, this symptom is not catastrophic. The worst that can happen is dilution of the oil in the sump and hard starting.

Fig. 4-36 The effects of preignition. One can imagine the condition of the piston (Courtesy Champion Spark Plug Co.)

The plug in Fig. 4-38 looks awful, but is only suffering from scavenger deposits which have developed because of the chemical composition of the fuel. The color could range from egg-yellow to beige. Cleaning and regapping should restore its performance.

The last photograph of the series illustrates a very common malady. The plug is simply worn out, as indicated by the rounded contacts and the pronounced taper of the side electrode.

While a careful appraisal of the insulator tip can provide information on combustion chamber phenomena along with some insight into the electrical functioning of the plug, visual diagnosis is not infallible.

Spark plug reading is primarily concerned with combustion problems. You can get some idea of the electrical behavior of the plug, but the symptoms will be masked. For

Fig. 4-37 Carbon fouling—usually the result of worn rings (Courtesy Champion Spark Plug Co.)

Fig. 4-38 Scavenger deposits (Courtesy Champion Spark Plug Co.)

Fig. 4-39 Electrode erosion signifying the plug has reached the end of service (Courtesy Champion Spark Plug Co.)

example, misfiring under load will leave its mark (presumably) coming from incomplete combustion. The cause of the misfiring may be tracking (the plug fires from the insulator to the shell), an internally cracked insulator, or some electrochemical phenomenon such as lead fouling. Lead is an ingredient in most gasolines and can coat the insulator, giving the current a direct path to ground. It cannot be detected by the ordinary mortal eye, although experts claim they can see it as a slight gloss on the insulator.

Can small-engine spark plugs be cleaned? This question has exercised the minds of the some of the best engineers in the country. Spark plug manufacturers generally answer "yes." Briggs & Stratton and some 2-cycle builders emphatically say "no!" Their fear is that some of the abrasive will be left on the insulator and damage the engine.

Chapter 5
Carburetors & Fuel Systems

There is something about a carburetor which encourages tinkering. If the engine refuses to start, the temptation is to give the adjustment needles a few turns. The usual result is that the problem is compounded. Not only do we have an engine which will not start, we have one with a maladjusted carburetor as well.

Carburetors hold their adjustments over long periods. Some designs go so far as to have fixed jets so that the initial factory setting cannot be changed in the field. Nor do carburetors wear out in normal service. There are few moving parts and none of the parts are heavily stressed. There is some wear concentrated in the float valve (needle and seat assembly) and in the throttle shaft bearing bosses.

Most carburetor problems are the result of contaminated fuel. Rust particles, water, and gum residues in the gasoline supply eventually silt over the tiny passages and ports. If fresh fuel were used, and the owner drained the system during the off-season, the carburetor would rarely require maintenance.

Carburetors are complex devices which have multiple functions. When they fail, some intelligent troubleshooting is required to locate the problem. Besides being a waste of time and energy, uninformed tinkering can actually damage the instrument.

Carburetor Functions

For combustion to take place, gasoline must be mixed with oxygen and then ignited. A bit of gasoline in a thimble will burn sluggishly, producing a relatively cool, smoky flame. Only the upper surface of the liquid is exposed to oxygen-bearing air. If we take the same thimbleful of gasoline and break it into droplets that are exposed to oxygen on all sides, the mixture we have is a much more potent brew. In fact, when mixed with air to 2.25%, the explosive potential exceeds that of an equal weight of dynamite!

The primary function of the carburetor is to break the fuel into usable droplets. This process is called *atomization* and is achieved by jetting liquid gasoline into a high-speed stream of air. Once in the engine, the fuel droplets go through a further change and become vapor.

The second function is to feed the engine the correct mixture of fuel and air. The amount of power produced by an engine is, in the last analysis, a function of the amount of air induced. It takes about 7 lb of air to produce 1 horsepower for one hour. This figure varies somewhat with engine design, but it is reasonably accurate. The horsepower output of a given engine fluctuates with speed and load. The carburetor automatically regulates fuel delivery according to horsepower requirements.

At light-load cruising speed, most small engines are happiest with a mixture of about 16.5 parts air (by volume) to 1 part gasoline. At high speed, the mixture must be richer and the proportion of air drops to 13 or 14 to 1. Because engines do not idle efficiently, an even more rich fuel mixture is needed during idle—the proportion may be on the order of 10 to 1. The richest mixture occurs during cold starts. When the engine is cold, vaporization will not take place. Only a trace of the gasoline blend boils (commercial gasolines contain a few aromatics which boil at temperatures as low as −40°F). An

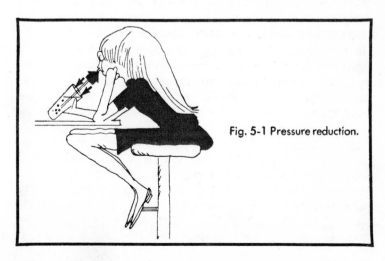

Fig. 5-1 Pressure reduction.

extremely rich mixture is required to pump enough of these aromatics into the engine for starting.

During the course of engine operation the carburetor must provide a range of mixtures—very rich for starting, rich for idle, relatively lean for cruise, and rich again at full throttle under full load.

The third major function of the carburetor is to control the speed of the engine by varying the total air/fuel input. Unlike auto or aircraft engines, the lawnmower throttle position is not entirely up to the whim of the user. A lawnmower throttle has an override in the form of a centrifugal governor. The user cannot "overrev" the engine under light loads, and the throttle automatically opens to compensate for sudden loads.

CARBURETOR PRINCIPLES

All carburetors follow the same broad physical laws. Although individual designs differ in detail, if you understand one carburetor, you understand them all.

The basic principle involved is that fluid flows in a direction to equalize pressures. Look at Fig. 5-1 for a moment. By drawing, the girl causes a pressure drop in the straw. The fluid in the glass responds to this pressure drop by moving up the straw. It is literally pushed up the straw by air pressing down on the surface of the liquid in the glass. The amount of fluid which will flow during any given period depends primarily upon two factors: the degree of pressure differential and the inside diameter of the straw.

The weight of air at sea level produces an atmospheric pressure of 14.7 psi (pounds per square inch). This pressure is termed 1 atmosphere. We measure subatmospheric pressures in terms of torr (meaning millimeters of mercury, and abreviated T). Because it has a very low melting point, mercury remains in the liquid state at low temperatures. If we pour some mercury in an open bowl and insert a tube vertically into the liquid, the levels of mercury in the tube and the bowl will be the same. Now, if we connect the end of the tube to a vacuum pump, the pressure differential will cause the mercury to rise (Fig. 5-2). How far the column rises is a measure of the vacuum created by the pump.

If our pump could generate a perfect vacuum, the mercury would rise to 748 mm (29.92 inches). Of course it is impossible to create a perfect vacuum on this planet. We have to journey to outer space for that.

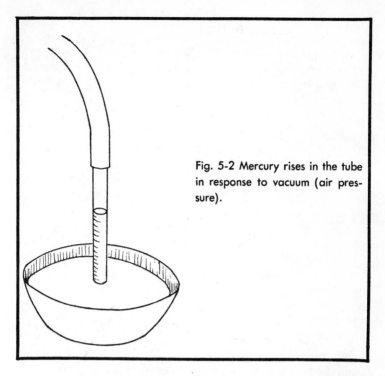

Fig. 5-2 Mercury rises in the tube in response to vacuum (air pressure).

In a previous discussion a gasoline engine was described as being like a pump during the intake stroke. Manifold vacuum varies with engine design, but generally hovers around 450T at idle. Gasoline in the carburetor is at atmospheric pressure. The pressure difference caused by manifold vacuum makes the fuel move through the carburetor.

However, manifold vacuum is not entirely adequate to insure consistent fuel delivery. The vacuum generated on the intake stroke depends upon the position of the throttle valve. When the valve is closed, or nearly closed, the engine is in effect pulling on a plugged pipe—vacuum readings are high. But when the throttle is opened, the engine attempts to inhale the whole atmosphere—vacuum plummets; fuel delivery becomes erratic and, as pressures equalize, eventually stops.

We need some method of boosting manifold vacuum at cruise and full throttle. Fortunately, Daniel Bernoulli discovered a simple way to do this back in 1775, long before gasoline engines were even dreamed of. He invented the *venturi*.

The venturi is a vacuum pump without moving parts, being merely a restriction placed in the carburetor bore. As you can see in Fig. 5-3, the restriction is gently curved to reduce turbulence and (consequently) inefficiency. However, any restriction will work.

The moving air stream is squeezed by the venturi. As a result, its velocity increases and its pressure drops. The numerical values in Fig. 5-3 are purely arbitrary, but they illustrate the principle. Every venturi operates on a constant devired by multiplying air speed by pressure. This figure never changes. In the example, the air speed at the mouth of the carburetor is 5 and the pressure is 4; therefore, the constant is 5 × 4, or 20. At the restriction, the air speed has doubled to a figure of 10, and pressure drops to 2. The constant (10 × 2) still holds. Downwind, the figures return to their initial values. The constant represents the sum of energy in the system. Since energy cannot be created or destroyed, the venturi trades off pressure for velocity.

Another way of understanding the principle is to consider that the amount of air leaving the carburetor bore must be the same as the amount which enters. When the air stream encounters the restriction we must still get the same amount of air through a smaller diameter opening in the same time period. Consequently, velocity increases. The law of conservation of energy dictates that a velocity increase follows a pressure decrease. Fuel enters the bore at the point of maximum velocity and lowest pressure. The tube shown in Fig. 5-3 is the fuel outlet nozzle; it is placed at the point of maximum velocity and lowest pressure.

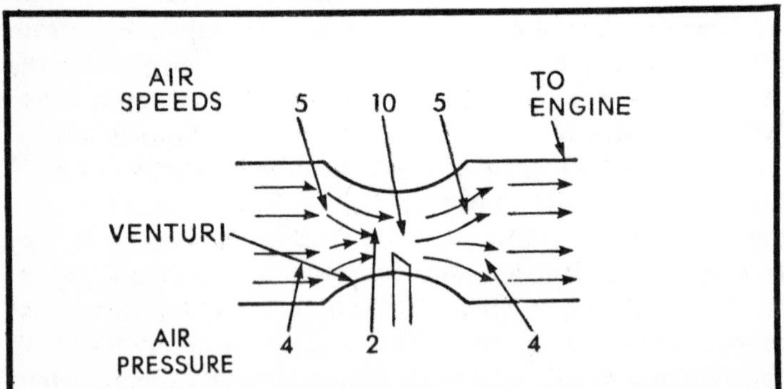

Fig. 5-3 Venturi action.

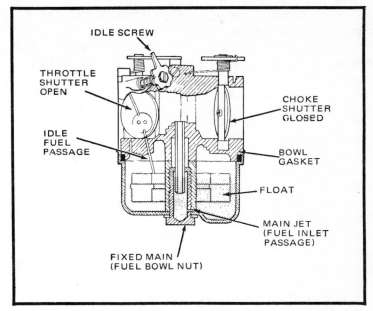

Fig. 5-4 Side-draft carburetor in cutaway view (Courtesy Tecumseh Products Co.)

The area of the bore is quite critical. At wide throttle, the area determines the amount of fuel and air induced. In this sense, bore area determines the horsepower potential. But we cannot increase the area beyond certain fixed limits, for as the area goes up, the air stream velocity drops.

Because the best fuel atomization and the greatest power occurs when air stream velocities are in the range of 300 ft/sec, the bore must be somewhat smaller than the ideal. And since air velocity is also a function of piston diameter and speed, we have to decide on a bore area for the operation speed range of the particular engine. Lawnmower carburetors have bore diameters of from 1 to 1½ in. to produce maximum power at 3200–3600 rpm. Switching carburetor bore diameters will drastically change the characteristics of the engine.

If the air stream through the bore is on the horizontal plane of the combustion chamber, we call the carburetor a side-draft type. Figure 5-4 illustrates such a carburetor. The major advantage here is compactness. A secondary consideration is that this configuration offers the most direct route to the chamber. Direct access means that pumping losses are reduced.

Updraft carburetors are used on a number of Tecumseh and Briggs & Stratton engines (Fig. 5-5). On horizontal crankshaft models the updraft configuration takes up little space and, in the event of flooding, isolates the chamber from liquid gasoline. Downdraft carburetors are found on a few of the heavier commercial machines. Figure 5-6 illustrates a design used on certain large Kohler engines.

Main Jet

The main jet—sometimes known as the *power* or *high-speed* jet—is mounted low in the fuel nozzle. A fixed main jet is shown in the cutaway view of the Tecumseh design in Fig. 5-4; an adjustable jet can be seen in Fig. 5-5. The fixed jet may be integral with the nozzle, in which case it is merely a drilled hole. More sophisticated carburetors have replaceable main jets which thread into the nozzle end, and are available in different orifice diameters. An engine which is operated at its power peak for long periods will run cooler with a larger-than-stock orifice. Engines operated at high altitudes can benefit from a smaller orifice; at high altitudes atmospheric pressure is lower, the charge contains less oxygen, so standard jetting would cause an overly rich condition. In general, flow should be reduced 2% per each 1000 ft of elevation above sea level.

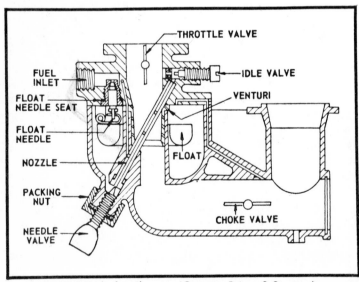

Fig. 5-5 Updraft carburetor (Courtesy Briggs & Stratton)

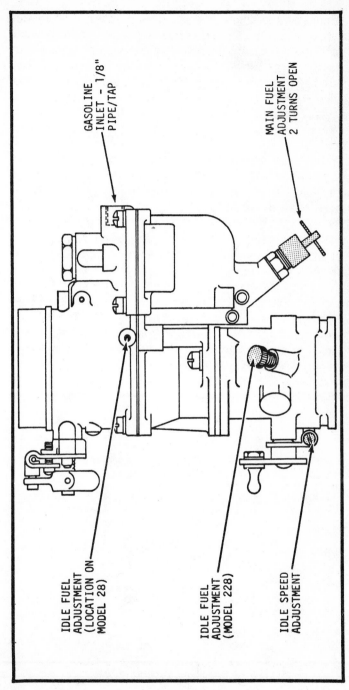

GASOLINE
INLET – 1/8"
PIPE/TAP

MAIN FUEL
ADJUSTMENT
2 TURNS OPEN

IDLE FUEL
ADJUSTMENT
(LOCATION ON
MODEL 28)

IDLE FUEL
ADJUSTMENT
(MODEL 228)

IDLE SPEED
ADJUSTMENT

Fig. 5-6 Downdraft carburetor (Courtesy Kohler of Kohler)

Of course, the adjustable jet allows variations to match operating conditions. Since no two engines are exactly alike, an average factory calibration is never completely satisfactory. One disadvantage of the adjustable jet has already been alluded to—it encourages tinkering. Another is that if the needle is screwed down hard against the orifice taper, the soft brass parts will distort; even the slightest distortion can upset the mixture enough to cause uneven running.

The fuel nozzle may have a series of radial holes drilled near the outlet (Fig. 5-7). These holes form an air-bleed system, the principle of which is shown in Fig. 5-8. Without an air bleed, the soda would drop back down the straw as soon as suction was released. In terms of a carburetor, if the throttle were to open suddenly, the fuel would drop down the nozzle and the engine would momentarily starve. The middle drawing illustrates a simple air bleed. The hole in the straw breaks the soda into bubbles which tend to cling to the walls. The last drawing is a refinement of the principle. The hole is located below the soda level and feeds air through a pipe. The bubbles are smaller and easier to atomize. In addition to holding the fuel level constant and atomizing the fuel, the air bleed provision prevents siphoning during shutdown.

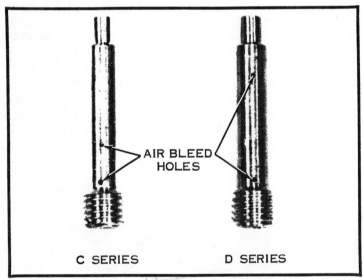

Fig. 5-7 Air bleed holes (Courtesy OMC)

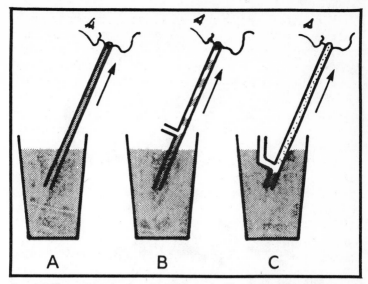

Fig. 5-8 The air bleed principle. Without the bleed, liquid would fall back down the straw when pressures equalized (A) A bleed port above the liquid level helps atomize the fluid and holds it in the straw (B) A bleed under the fluid level does a much more thorough job of atomization (C).

An air bleed can also be used to limit fuel delivery at high speed. As piston speed goes up, the air velocity through the venturi increases. Since the venturi is a creature of velocity, it responds by creating a further pressure drop and draws in more fuel. Up to a point this is good, since the engine's appetite grows with speed. But there is a point at which the venturi delivers too much fuel. To compensate for this phenomenon, engineers often place the main-jet air-bleed inlet in the air horn. As velocity increases, the inlet scoops more and more air and dumps it into the main jet or its environs. At very high speeds, this air forms a barrier, preventing fuel from passing through the jet, and causing the mixture to become leaner in a ratio that the engine can tolerate.

At nearly full throttle the carburetor works as shown in Fig. 5-9. This Tecumseh design is very typical of lawnmower carburetors in that some fuel flows through the low-speed jets even with a fully opened throttle.

Low Speed

The throttle valve is downwind of the venturi. At idle and off-idle the valve (variously known as the damper, disc,

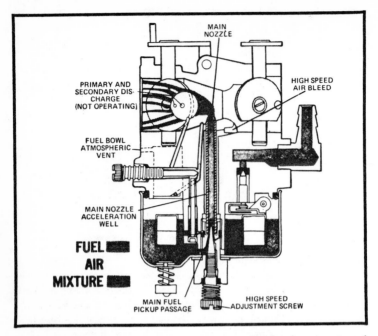

Fig. 5-9 High-speed operation. Suction is provided by the venturi, and fuel is lifted through the main jet. Note the provision for air correction labeled HIGH SPEED AIR BLEED (Courtesy Tecumseh Products Co.)

butterfly—and labeled shutter in Fig. 5-10) effectively shrouds the venturi. Air velocity drops so much that no fuel can pass through main nozzle. Instead, fuel enters through the low speed jets. The term "jets" may be a trifle misleading in this context. We usually think of a jet as a distinct and removable part. The idle and off-idle jets are actually ports that open into the sides of the carburetor bore. The idle jet is located directly below the valve. One or more off-idle jets are stacked directly above it. Normally, these jets are round, although there has been a tendency in recent designs to elongate them.

The edge of the throttle disc almost touches the inside wall of the bore. It acts as a venturi, forcing the air stream to speed up and lose pressure. Consequently, the idle jet flows. You will note in Fig. 5-11 that during idle, the off-idle jet acts as an air bleed. As the throttle swings wider, the off-idle jet begins to pass gasoline. Figure 5-12 illustrates both of these jets—labeled primary and secondary idle discharges—in operation.

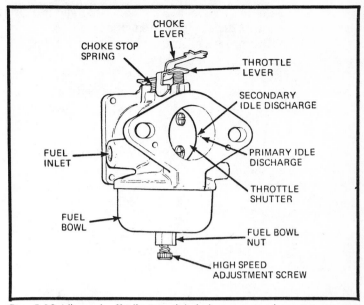

Fig. 5-10 Idle and off-idle jets (labeled PRIMARY and SECONDARY IDLE DISCHARGE) (Courtesy Tecumseh Products Co.)

Because gasoline engines are supersensitive to mixture variations at idle, most carburetors have an idle-adjust needle. Typically, the needle restricts gasoline flow through the jet as shown in Fig. 5-12. To make the mixture richer, turn the needle counterclockwise so that it retracts out of the jet. You may run into a carburetor which adjusts low speed mixture at the air bleed. This is typical European and Japanese practice and can be found on Amal, Bing, and Keihin designs as well as on certain American models. Both Bendix and Tillotson (MD

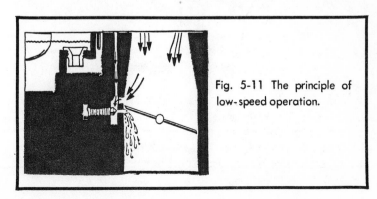

Fig. 5-11 The principle of low-speed operation.

series) have used this approach. Tightening the idle mixture screw enriches the mixture by restricting the air bleed.

The only difference between the two, from a mechanic's point of view, is that the air-adjust type becomes richer as the needle is turned inward. Air-adjust needles are typically blunt and rounded as opposed to the sharply pointed fuel needles.

Most carburetors have a third external adjustment in the form of an idle speed regulating screw, used to limit throttle travel.

As a rule, the low-speed mixture-adjustment needle is located near the throttle plate. The high-speed needle is below the float (by far the most common location) or on the side of the bore. Carter and a number of OMC designs may have this needle positioned vertically above the bowl. The idle speed regulating screw is generally at the throttle plate shaft. A few models have the regulating screw on the side of the bore where it might be confused with an idle mixture adjustment.

Starting Circuit

After some experimentation, American designers have settled upon a choke valve (or plate) to provide the rich

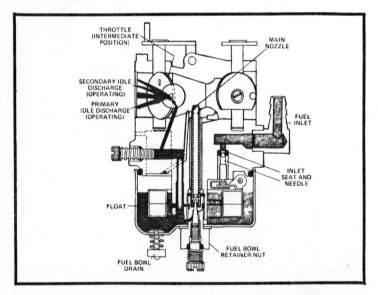

Fig. 5-12 Off-idle, or intermediate operation, throttle shutter "cracks" to decrease restriction. Engine runs on medium fuel air mixture (Courtesy Tecumseh Products Co.)

CHOKE PLATE

Fig. 5-13 Choke cutaway to provide some air when the valve is closed (Courtesy Tecumseh Products Co.)

mixtures needed for starting. The valve is mounted upwind of the venturi, directly behind the air filter, and resembles a throttle valve. When the choke is closed, the carburetor bore is subject to manifold vacuum. Consequently, idle jets, off-idle jets, and main jets flow.

Some chokes have one edge trimmed (Fig. 5-13) or feature a tiny poppet valve (Fig. 5-14). These expedients help prevent flooding by diluting the mixture with air.

Foreign engines (which are being imported in ever increasing numbers) sometimes have chokeless carburetors; the mixture is enriched by means of a large-diameter starting jet. The advantages claimed are less restriction of the throttle bore and a reduced tendency to flood.

Several Tecumseh and OMC designs used a priming pump in addition to the conventional choke. If operated properly, the pump can help insure one-pull starts by injecting fuel directly into the carburetor throat.

Internal Fuel Level Regulation

The standing fuel level in the main nozzle and in the idle circuit passageways determine, in part, the relative richness of the mixture. The level must be controlled over narrow limits, regardless of the amount of fuel in the tank or the pressure produced by the fuel pump.

Most small-engine (and automobile) carburetors use a float and valve assembly to control the fuel level. The float pivots upward on a pin and closes the inlet valve when the fuel in the chamber reaches a predetermined level. Modern floats are made of injection-molded plastic or of sheet brass crimped and soldered into a sealed structure. A few carburetors still use cork floats which have been varnished to prevent saturation. These floats have almost disappeared because of the difficulty in obtaining consistent results with organic materials—cork varies in density from sample to sample.

Various float styles have been used on lawnmower engines. Perfectly spherical floats are of historical interest. Marvel-Schrieber carburetors, fitted to Power Products engines, had disc-shaped floats. The current practice is to employ a doughnut float as shown in cutaway in several illustrations in this chapter.

A concentric main jet feeding through the float remains covered with fuel at all but the most extreme angles of tilt. Thus, the user does not have to concern himself with fuel starvation (or the other extreme—flooding) when mowing on hillsides.

The inlet valve, or more properly, the inlet needle and seat, has been subject to intense development. It is the single most troublesome part of a conventional carburetor, and the one with the most potential for injury. If it should stick open, the carburetor will flood with raw gasoline. This represents a

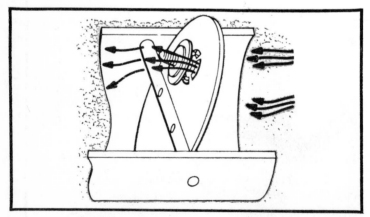

Fig. 5-14 Choke poppet valve. This nicety has all but disappeared from contemporary carburetors, but you will encounter it in older Carters.

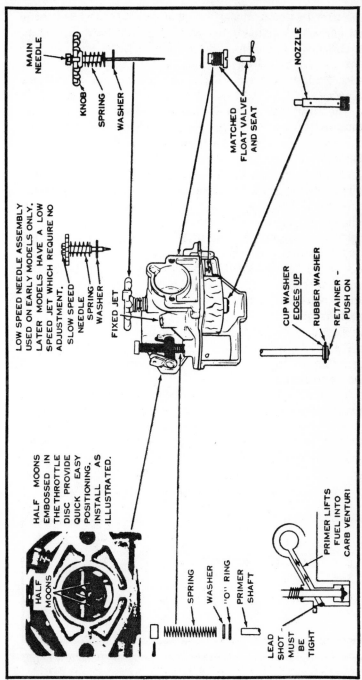

Fig. 5-15 Outboard Marine carburetor design with a primer (Courtesy OMC)

139

fire hazard as most modern engines are built on the **L**-head configuration, with the intake and exhaust valves working off one camshaft on the same side of the bore. The hazard is made more serious by the increasingly popular use of plastic fuel tanks. Should the valve stick closed, the engine will simply starve.

The valve seat is usually made of brass, although more exotic materials are gaining on this old standby. Tecumseh favors a synthetic, Viton (a trade name), for seat material. Interestingly enough, certain Rochester Quadrajets use a similar synthetic. The needle—and here we might generate a bit of confusion since the term "needle" is also used for the mixture adjustment screws—is normally made of high-nickel-content steel. The combination of brass, which is a relatively soft metal, and stainless steel guarantees long life. Briggs & Stratton has had much success with plastic needles working against brass. Other manufacturers coat the tip with a rubber compound. This material is soft enough to conform to minor imperfections on the seating surface and can tolerate small rust particles..

Most needles have a provision for positive opening. A clip spring, or an eyelet, on the needle ties it to the float as shown in Fig. 5-16. The needle may also incorporate a spring-loaded damper to muffle float action.

The float bowl is usually in the form of a steel stamping, galvanized on the inner surface to forestall rust. The bowl may have a fuel drain valve. Although troublesome in the sense that these drains develop leaks as the rubber sealing washer ages, this feature is, on the balance, a good one. It enables the owner to drain the fuel system prior to storage and is a convenient check point during fuel system diagnosis.

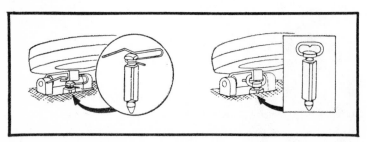

Fig. 5-16 Needle variations to assure positive opening (Courtesy Briggs & Stratton)

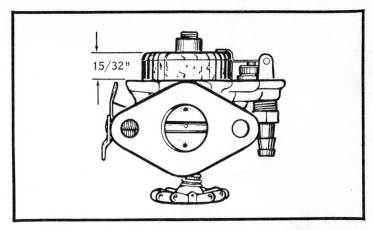

Fig. 5-17 Float height adjustment. The dimension shown is typical, and does not apply to all carburetor types, or to different engine applications of a single type (Courtesy OMC)

The bowl must be vented to the atmosphere; otherwise, the engine will stop as soon as float bowl pressure drops to the value at the venturi. Many designs are vented at some point on the external surface of the main body casting. The disadvantage of this approach is that the vent can easily clog in lawnmower and edger service. The combination of dust and fuel seepage through the pot-metal castings makes the vent very vulnerable. A better solution is to vent the chamber into the carburetor throat.

Float height is a most critical adjustment. Adjusted too low, the carburetor turns rich—too high, and the opposite condition prevails. The exact adjustment varies with different makes and models. The basic idea is to assemble the float and needle-and-seat, turn the carburetor casting upside down, and measure the distance between the float and the chamber roof. The float tang—the brass strip between the float proper and the needle—is bent with a pair of long-nose pliers. Apply minimal force to the needle, especially if it has a soft tip.

Although concentric fuel pickups help, extreme engine tilt and vibration can still affect mixture consistency; the float has inertia which limits its sensitivity of response. A typical needle-and-seat reciprocates at about 200 cycles per minute. A decade ago, it appeared that the diaphragm carburetor offered the best solution. Instead of a cumbersome float, a Neophrene diaphragm regulated fuel delivery by demand from the engine. Response was keyed to engine speed and, at wide-open throttle, could approach 4000 cycles per minute. These carburetors were not dependent upon gravity, as the concept was first developed for acrobatic airplanes. Since response was so sensitive, very little fuel remained in the carburetor reservoir after shutdown, and one could reasonably expect fewer gum and varnish problems. These units were quite compact—a factor which reduced the overall size of the package, simplified mounting problems, and cut shipping costs. All in all, the prospects looked rosy.

However, once in service, these carburetors developed problems. Some of the difficulties could be traced to design inadequacies, especially for lawnmower applications. Perhaps the biggest problem was the attitude of the service technicians. Mechanics are a notoriously conservative lot. They dislike design innovations as a rule. Each innovation brings with it a period of trial and error for the mechanic and, in a sense, makes some of his own accumulated skill obsolete; engine manufacturers seldom instituted training programs to acquaint mechanics with new designs, even though diaphragm-pot theory could have been taught in an hour or so. Repair costs reflected the mechanic's ignorance and frustration. Customers were warned never to purchase an engine with one of those funny-looking carburetors on it.

Other problems involved the diaphragm and needle-and-seat. These parts were quite delicate. The diaphragm would deteriorate if gasoline was left in the tank for extended periods. In many cases, the diaphragm would fail without giving any visual clues that anything was wrong. A few hundredths-of-an-inch of stretch seemed to be enough to cause it to flutter uselessly. The needle-and-seat mechanism was spring-loaded in the closed position, and tended to remain closed after storage. Careless disassembly often bent the needle beyond usefulness.

We will discuss the operation and repair procedures for a typical diaphragm carburetor in a later section. Many of these pots are in the field (some are still being fitted to lawnmower and edger engines), and as long as they are being properly maintained, should continue to give excellent service.

In summary, a carburetor consists of a venturi, a fuel nozzle in a main jet, a throttle plate, two or more low-speed jets, assorted air bleeds, a choke plate, and either a diaphragm or a float to regulate the internal fuel level. With the exception of the venturi, any of these components are subject to failure.

GENERAL TROUBLESHOOTING PROCEDURES

Carburetors, relative to most other engine components, are reliable. Check the ignition and compression, before you assume the carburetor to be at fault. Do not change the mixture needle settings without good reason. We will go over the symptoms of carburetor failure and indicate the likely prognosis.

Fuel Starvation

If repeated cranking with the choke fully extended does not produce a wet spark plug tip, you can be sure that gasoline is not getting to the chamber. If this is the case, work from the tank forward; check the fuel level—you would be surprised to learn how many working mechanics crank a dry engine.

Move down the line to the carburetor; loosen the fuel inlet connection enough to determine if gasoline has reached that point. Engines with remote pumps will have to be cranked—ignition grounded to reduce fire hazard. Proceed to the float bowl. Walbro and Tecumseh carburetors have a drain valve; other designs must be partly disassembled.

The condition of the float chamber is an index of overall carburetor cleanliness. A layer of rust or chalky deposits indicate water in the system, including the tank. Gum and

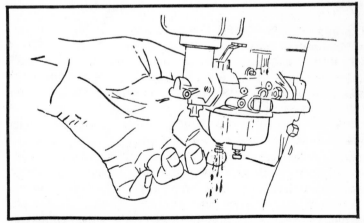

Fig. 5-18 Depressing drain valve to check fuel supply (Courtesy Clinton Engines Corp.)

varnish can be attributed to fuel left standing over long periods. In extreme cases, the tank and lines will also be contaminated.

More than a few hundredths-of-an-inch of sediment in the float chamber calls for a thorough cleaning of the carburetor and related components. As a short-term expedient, you may get started by blowing out the main discharge nozzle.

So far we have discussed fuel starvation in terms of stoppages somewhere in the system. Air leaks can produce the same result. Remember, a carburetor is a pressure differential device; vacuum must be maintained for it to function properly. Check the carburetor-to-block joint. Loose mounting screws or a severely blown gasket can leak enough air to prevent starting. Two-cycle engines can blow crankshaft seals, causing a loss in crankcase compression.

Persistent Flooding

Any carburetor will flood if taken through enough cranking cycles. Briggs & Stratton updraft types dribble gasoline on the slightest pretext. Flooding due to repeated cranking causes atomized fuel particles to collect on the walls of the intake tract and form puddles. Eventually, the spark plug tip will become wetted and will short, making starting impossible. In order of severity, the symptoms of flooding are:

- Fuel mist expelled out the exhaust pipe during cranking
- A wet spark plug insulator
- Raw fuel dribbling out the carburetor air horn

Persistent flooding can almost always be traced to the carburetor's internal fuel regulating mechanism, usually in the form of a malfunction at the inlet needle and seat. A speck of dirt or wear will cause the valve to leak. The float may also be at fault. Sheet brass floats can develop leaks with time. (You may be able to hear fuel sloshing inside the float.) A surer test is to boil the float in a tin of water; the heat will open any pin holes and the float will sink. Never attempt to repair a float. However, a few grams of solder will increase the weight of the float enough to richen the mixture all through the range. The varnish coating on cork floats can crack, allowing the cork to absorb fuel. Another point to check is the pivot and pin. Pronounced wear can, in some designs, allow the float to cant and hang on the sides of the chamber. Some carburetors have welch or expansion plugs located under the float or under the diaphragm chamber fuel level. These plugs are a manufacturing convenience and are not designed to be removed in normal service operations. Should they vibrate loose, excessive fuel will be delivered to the jets. If new plugs are not available—and they rarely are—the best bet is to epoxy the originals into place. Scrape the interface and clean with lacquer thinner. Use 24-hour epoxy, since it is stronger than the quick-curing varieties.

Another fairly common cause of flooding is a choke valve which remains partially closed. These valves are spring-loaded to stay, but heavy dirt accumulations, and the paint on new mower shafts, can cause binding. It is usually not necessary to disassemble the valve and shaft—lacquer thinner applied with an eyedropper to the shaft bearings will do the trick.

A main jet which is way out of adjustment will also cause flooding by delivering a mixture too rich for hot starting. In broad and general terms, the engine should run with the main needle seated finger-tight and backed off 1¼ turns. We will be more specific on this point in the following section on individual carburetors. As a final measure, inspect the air filter; clean, as necessary.

Gas Starvation

An engine which starts but shows symptoms of gas hunger at certain speeds is generally suffering from an intake malady. Lean running can be a difficult ailment for the amateur to diagnose. In outline, the symptoms are:

- Excessive chamber temperature shown by spark plug condition (see the following chapter)
- Lack of power
- Lean roll (a drop in rpm at a set throttle position)
- Spit-back through the carburetor throat (may be initiated by opening the throttle quickly)

Assuming that the adjustment is correct, two conditions are the cause of lean running; an air leak in the intake tract is the most common and tends to be more pronounced at idle. Check the carburetor mounting flange, the cylinder head gasket and, on 2-cycles, the crankshaft seals. Fuel stoppages can also produce the same symptoms. Generalized lean running throughout the speed range is a manifestation of insufficient fuel delivery, originating either in the tank plumbing or in the main fuel nozzle. Some carburetors feed all systems through the main nozzle and others have a separate idle-speed pickup (the main nozzle cuts in at fairly low speeds). Refusal to idle typifies a stoppage in the idle jet or off-idle jet. Diaphragm carburetors are subject to these particular maladies as well as to direct diaphragm failure.

Gas Gluttony

The symptoms of this malady are:

- Black, sooty spark plug tips. (The inside of the exhaust pipe may be coated with carbon fluff as well.)
- Black smoke accompanied by the odor of gasoline
- Lack of power
- Four-stroking in 2-cycles. (The engine will fire every second revolution with a distinct pop.)

Rich running is almost always a matter of carburetor adjustment; always check the float setting first. Other possibilities are a sticking choke, a clogged air filter, or (and this is a rare one) lubricating oil contaminated with fuel. Engine makers sometimes vent the crankcase into the carburetor air horn. Besides assuring positive ventilation, this approach helps keep the mower clean and cuts emissions as well. Crankcase vapors are said to account for a quarter of all HC (hydrocarbon) emissions. In very old engines with loose pistons, enough gasoline can leak by to make, in effect, a second fuel supply.

A more likely cause of rich running is a clogged air filter element. (We will go into filters in more detail later in this chapter.) It is enough to say here that the filter can be considered to be clogged if removing it makes a *marked* difference in engine performance. Some lawnmower filters are restrictive by their very nature, but a clean filter should not produce more than a 200 rpm change.

Poor Tune

If you have checked the compression and the ignition, and replaced the spark plug, but are still plagued by ragged, uneven running, clean and overhaul the carburetor. Replace the wearing parts and inspect the adjustment needles carefully. If they are grooved or eroded, throw the rascals out. Grooves will not stop the engine from running, but they will make precise adjustment impossible. Do not attempt to file or otherwise reshape the tips.

GENERAL SERVICE PROCEDURES

Once you have determined to go operate on the carburetor, the first thing to do is to clean the unit without removing it from the motor. Kerosene and Gunk followed by a hosing should do the trick. A word of caution before hosing: **Remove paper element air filters and keep water from the exhaust pipe and flywheel.** Be sure to note the position of the linkages and the spring attachment points. Disconnect the fuel line (a pencil serves as a stopper for synthetic lines) and remove the mounting flange screws. Pull the carburetor free at the flange and work the throttle cable free, being careful not to bend the cable or the linkages in the process. Most throttle assemblies incorporate a magneto kill switch which will have to be disconnected. A few models have tank-mounted controls which means that you will have to remove the shrouding and manual starter as one assembly.

Now, with the carburetor in hand, move to a clean spot on the bench and begin disassembly. As you do so, you will find that lawnmower carburetors are actually quite simple in construction.

For cleaning, most shops prefer Bendix Econo-Cleane, although there are other good chemical cleaners on the market. Delco X-55 does an excellent job and is water-soluble; other preparations must be removed with kerosene or Varsol. Unfortunately, most carburetor cleaners are sold in 5-gallon tins. Gunk carburetor cleaner (not to be confused with the degreaser under the same trade name) comes in pints. It is not as potent as some of the commercial products, but it is adequate.

Use caution around any of these cleaners. In addition to gum, varnish, carbon, and paint, *they will remove skin.* Do not inhale the fumes, and certainly don't smoke around an open container. Suspend the parts to be cleaned in a tray, and soak—but no longer than necessary. Although claimed to be harmless to metal, cleaners will give a stippled effect to castings if the exposure is excessive.

Disassemble the carburetor only so far as necessary to remove all rubber and synthetic parts. It is not necessary to disassemble the choke and throttle discs, or the welch plugs. (Some manufacturers suggest the plugs be removed, but before you do so, make certain your dealer has replacements in stock.)

After half an hour or so, the carburetor should be reasonably clean. In extreme cases—water damage and the like—you may have to soak it longer. Flush the cleaner and inspect each part with a gimlet eye. Replace gaskets, O-rings, diaphragms, and needle-and-seat assemblies as a matter of course. Set the float level to specifications and assemble. Make the preliminary mixture adjustments and mount the carburetor on the mower. Adjust the linkage so that the choke closes completely at one end of travel, and the ignition switch closes at the other. You may have to do some judicious bending to satisfy both of these requirements.

Final mixture adjustment varies between different makes and models. In general, follow these steps:

1. Fill the tank to the halfway mark with clean fuel. Rust or water-clouded gasoline will send you right back to the bench.
2. Run the engine for 10 or 15 minutes in a well ventilated place to let it reach a stable operating temperature.
3. Rotate the screws $1/16$ turn at a time, allowing a few seconds between adjustments.
4. Begin with the main jet needle. At ½ to ¾ throttle, turn the needle inward to lean roll; then back it out to rich roll, and split the difference.
5. Turn the low-speed screw for best (fastest) idle.
6. Reset the high-speed screw. The systems overlap; one adjustment affects the other.
7. From idle, tap the throttle. The engine should accelerate cleanly without stumbling. If it hesitates, open the main needle a fraction of a turn, and retest.
8. Set the idle speed regulating screw to the desired speed. Lawnmower engines should idle at 700−900 rpm.
9. Finally, check the main jet setting in the grass, under full load. You may find that the mixture should be enriched a fraction.

This procedure is really less complicated than the telling of it. An experienced mechanic can set a carburetor in 30 seconds or less.

SUCTION LIFT CARBURETORS

Suction lift carburetors are a class unique to lawnmowers and edgers. They are primitively simple—so simple, in fact, that some mechanics refuse to call them carburetors and use

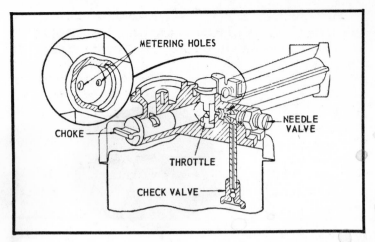

Fig. 5-19 The "Vacu-Jet" suction lift carburetor (Courtesy Briggs & Stratton)

the term "mixing valve" instead. Yet, simplicity has certain advantages, especially in terms of reliability—nonexistent parts cannot fail.

Briggs & Stratton

The *Vacu-Jet* is the most popular suction-lift carburetor. It has been used on millions of reel-type mowers and edgers in several variations.

The main body is a pot-metal casting, and contains the high- and low-speed jets (identified as "metering holes" in the cutaway drawing in Fig. 5-19). The high-speed side has an adjustment needle. The pickup tube is fitted with a ball check valve which traps fuel high in the tube for better responsiveness and total delivery limiting. If this ball sticks open, the engine will run rich. It is much more usual for the ball to stick in the down, or closed, position. It can be freed by inserting a thin wire from the bottom through the screen. Early-model tubes were made of brass and could be cleaned in the Bendix compound mentioned earlier. All later types have nylon tubes and balls.

The tubes are pressed into the body casting. Removal of the nylon types can be accomplished with a wrench on the hexagonal head. Pry the older ones free in a vise as shown in Fig. 5-20. Installed height should be no more than a few thousandths of an inch greater than ¼ inch on both models.

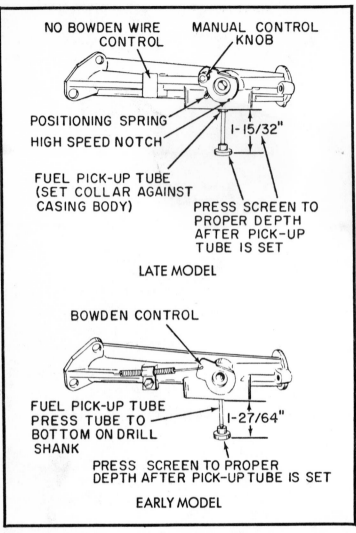

NO BOWDEN WIRE CONTROL

MANUAL CONTROL KNOB

POSITIONING SPRING

HIGH SPEED NOTCH

FUEL PICK-UP TUBE (SET COLLAR AGAINST CASING BODY)

I-15/32"

PRESS SCREEN TO PROPER DEPTH AFTER PICK-UP TUBE IS SET

LATE MODEL

BOWDEN CONTROL

FUEL PICK-UP TUBE PRESS TUBE TO BOTTOM ON DRILL SHANK

I-27/64"

PRESS SCREEN TO PROPER DEPTH AFTER PICK-UP TUBE IS SET

EARLY MODEL

Fig. 5-20 Suction-lift carburetors found on Lauson engines (Courtesy Tecumseh Products Co.)

Should the boss be loose, secure the pipe with Loc-Tite sealant, or hot fuel-proof model airplane dope.

These carburetors are prone to leakage at the tank flange. This condition is dangerous, since the tank is in such close proximity to the exhaust pipe end. Check both parts for trueness with a straightedge. Mill the faces with emery cloth

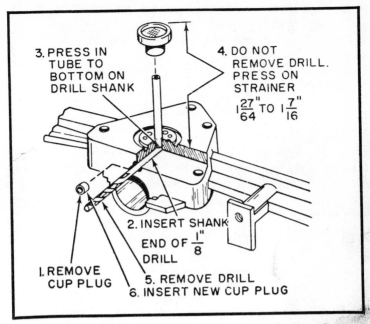

3. PRESS IN TUBE TO BOTTOM ON DRILL SHANK

4. DO NOT REMOVE DRILL. PRESS ON STRAINER $1\frac{27}{64}''$ TO $1\frac{7}{16}''$

2. INSERT SHANK END OF $\frac{1''}{8}$ DRILL

1. REMOVE CUP PLUG

5. REMOVE DRILL
6. INSERT NEW CUP PLUG

Fig. 5-21 Setting pickup tube depth on a Tecumseh carburetor (Courtesy Tecumseh Productions Co.)

as described under **Cylinder Heads** in Chapter 8. Look at the mounting bosses carefully for signs of cracking.

Assemble with new gaskets at the tank mounting and carburetor-to-block flanges. Initial main needle adjustment is 1¼ turns open from finger-tight. Run the engine under load to make the final setting. Chances are you will find that maximum power and best idle are contradictory. Set the idle fast—1750 rpm—and let it go at that. Fluctuating at light load is also a characteristic of this design and cannot be eliminated, short of reengineering the governor.

Tecumseh

The Tecumseh suction lift carburetor has been fitted to a substantial number of Craftsman rotary mowers. Two models are depicted in Fig. 5-20. This carburetor is distinguished by having no adjustments other than the control cable placement. The only tricky one is the older model. When installing a new strainer or tube, use a ⅛ in. drill bit as a depth gage (Fig. 5-21). Secure the cup plug with sealant. The new model has a reservoir cup which fits over the tube and into the tank. The slotted end of the reservoir must be down to keep the engine from starving.

Clinton

Clinton's suction lift is a fixed jet design, made from welded sheet steel. There are no special servicing instructions peculiar to the Clinton configuration (Fig. 5-22).

FLOAT MODELS

Briggs & Stratton's updraft design comes in several bore diameters. Known as the *Two-Piece Flo-Jet*, it is one of the most sophisticated carburetors used on small engines. It's also one of the easiest to tune. Refer back to Fig. 5-5 for a sectional view of this carburetor.

Disassembly entails unscrewing the packing nut and the main adjustment needle. The fuel nozzle enters a recess in the

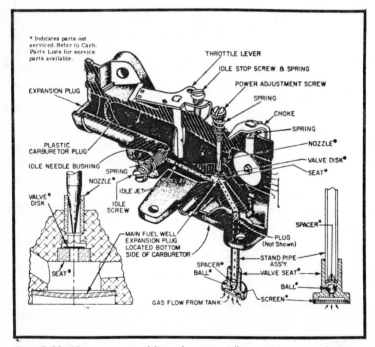

Fig. 5-22 Clinton suction lift carburetor. Idle setting varies between models. Those with an idle bushing (see inset) should be initially set at 4 or 4½ turns from seated. Carburetors which do not have this modification can be identified by the coarse-thread needle. Preliminary adjustment is 1 or 1½ turns open (Courtesy Clinton Engines Corp.)

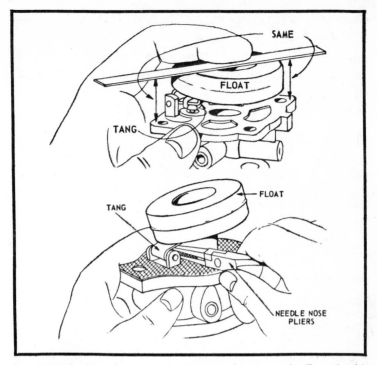

Fig. 5-23 Checking float level on Briggs' carburetors—the float should be parallel to the casting (Courtesy Briggs & Stratton)

throttle body and must be removed before the carburetor halves can be separated. Using B&S special tool 19061 or 19062, remove the nozzle from below. (You can make an equivalent tool by grinding the flanks of a blunt-tipped screwdriver.) Next, remove the four screws holding the two castings together.

The needle-and-seat assembly should be replaced if it is of the standard type. Late-production Flo-Jets are sometimes furnished with Viton seats. Customarily, these are left in place; but if you notice wear, or if the carburetor tends to flood, remove the old seat with a self-tapping screw. Install a new one, using the old seat as a buffer; the seat should be exactly flush with the carburetor body.

The float should be level with the casting when inverted (Fig. 5-23). This is the standard Briggs practice; imitation by other manufacturers who specify some odd fraction of an inch would be welcome. Note how the tang is bent without putting force on the needle-and-seat.

These carburetors feature replaceable throttle shaft bushings. Extract the old ones with a ¼ in. tap. Press the new ones into place—no finish reaming is required. The throttle body should be flush with the lower casting to prevent air or fuel leaks. If it is warped, lightly tap the ears upward; the gasket will compensate for small irregularities.

Assembly is quite straightforward. The main nozzle goes in after the two halves have been joined. Initial adjustment is ¾ turn open on the low-speed needle, and 1½ turns on the high-speed needle. Tune as outlined previously. Idle speed should be 1750 rpm for the light-alloy block engines, and 1200 rpm for cast iron.

The *One-Piece Flo-Jet* is a side-draft design used primarily on rotary lawnmowers. It comes in two variations. The smaller model is essentially similar in appearance and construction, except that the main jet needle is located below the bowl.

There are no special tricks to servicing these carburetors. The only thing to watch is the nylon choke shaft. Nylon is an excellent bearing material in that it behaves as if it were self-lubricating. It does complicate carburetor servicing, nevertheless; chemical cleaners convert it into something resembling bubble gum. The best cleaning bet is to carefully lower the carburetor into the solvent to a level just below the choke shaft. You can remove the shaft, but this requires prying off the welch plug.

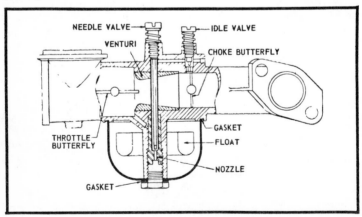

Fig. 5-24 One-piece Flo-Jet in cross section—smaller variant (Courtesy Briggs & Stratton)

The Walbro LM series encompasses a broad range of designs sharing basic castings and tooling. The number which you will find stamped on the air horn is keyed to detail

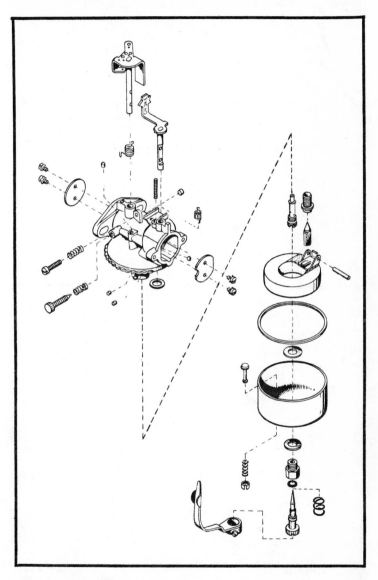

Fig. 5-25 Walbro LMG, LMB, and LMV series.(Courtesy Clinton Engines Corp.)

modifications in the control linkages and the like. These carburetors are found on Lauson (Power Products) and Clinton engines. They are fully adjustable and (except for gaskets) contain no plastic parts.

However, the LM carburetors have a peculiarity—the idle circuit feeds through a tiny cross-drilled hole in the threaded end of the nozzle (Fig. 5-26, part 23). Once the nozzle is turned, the idle passage is blocked. Consequently, the engine refuses to run at any speed below 2500 rpm. The annular groove allows fuel to pass, regardless of the nozzle position (Fig. 5-26). Although these nozzles are cheap enough, it is annoying to be forced to purchase a part. You can use the old one if you remove the tiny brass cup plug. Be careful not to drop the plug—it might be lost forever. Now, with a length of stiff wire, align the hole in the nozzle with the passage; drive the plug back into place.

The nozzle can also give problems if the carburetor has suffered water damage. The threads corrode, making removal difficult. Chase the threads with a ¼ in., No. 28 tap. Grind the end of the tap flat so it will bottom. If the nozzle still does not turn, remove it and install a new replacement.

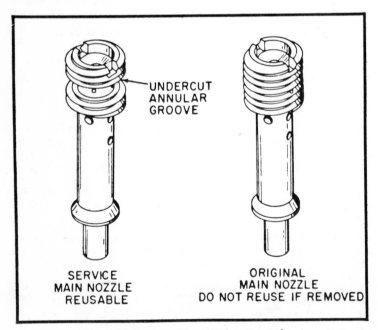

Fig. 5-26 Service and original main nozzle.

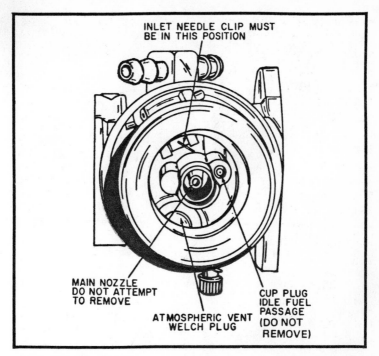

INLET NEEDLE CLIP MUST
BE IN THIS POSITION

MAIN NOZZLE
DO NOT ATTEMPT
TO REMOVE

ATMOSPHERIC VENT
WELCH PLUG

CUP PLUG
IDLE FUEL
PASSAGE
(DO NOT
REMOVE)

Fig. 5-27 Underside of the new Tecumseh carburetor. Unlike the Walbro, the main nozzle is not removable (Courtesy Tecumseh Products Co.)

LM series float are adjusted between the float and the main body as shown for the Briggs & Stratton. The exact setting varies between engine make and model. Consult your dealer for specifications.

Tecumseh has dropped the LM carburetors for a simpler design. Some of these have a provision for main jet mixture control, others are fixed. The major differences between the Tecumseh and the Walbro are the fixed main fuel nozzle, separate idle pickup, an internal float chamber vent, and a pressed in (rather than threaded) fuel inlet fitting (Fig. 5-27). Both feature a float chamber drain valve.

Two types of noninterchangeable float cover gaskets are used—square and round cross section. Pay particular attention to the needle clip position as a guide to assembly. If the fuel inlet fitting must be removed for any reason, install a replacement by inserting the tip into the boss and coating the external surfaces with Loc-Tite Grade A. Be sure the fitting is aligned properly for fuel line hookup.

The inlet seat, made of Viton, has been pressed into place. The easiest way to remove it is to put a few drops of oil into the cavity and connect an air hose to the inlet fitting. The seat will pop out like a champagne cork. If you do not have access to compressed air, you can extract the seat with a wire hook. The new seat should be driven in with a broad flat punch until it bottoms. Inspect the needle for wear, especially along the flanks where it rubs against the machines sides of the cavity. Float height varies with model and application, as does initial mixture setting. But 1¼ turns open on the needles should get you started.

Lawnboy uses one of the finest carburetors in the industry (Fig. 5-15). It is an adaptation of the OMC outboard designs. The main body is cast of aluminum (rather than pot metal) and has superior resistance to oxidation. As a bonus, the float bowl can be removed without tools.

Disassembly is straightforward. Remove the carburetor and the reed valve plate as a single unit. The cork floats should not be immersed in solvent. The float height setting given in Fig. 5-17 is for series A, C, and D engines. If the throttle and choke plate screws have been removed, replace them with slightly larger screws for security. Replace the Nyloc mounting nuts if the inserts show wear. A torn washer on D model primers will render the unit inoperative. Replace the whole assembly.

Early models with adjustable idle jets are initially set at one turn open. The main needle on all models should be opened two turns.

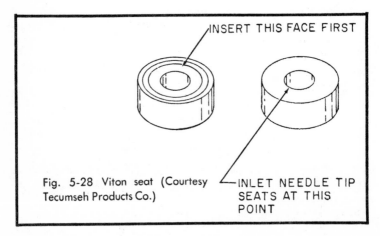

INSERT THIS FACE FIRST

Fig. 5-28 Viton seat (Courtesy Tecumseh Products Co.)

INLET NEEDLE TIP SEATS AT THIS POINT

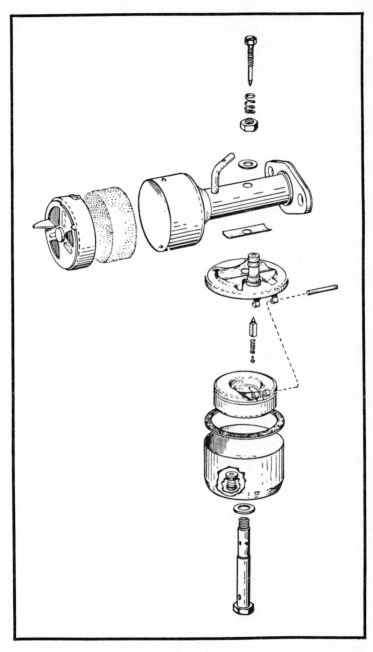

Fig. 5-29 A study in the art of the minimal—the Clinton 501 series made of sheet metal (Courtesy Clinton Engines Corp.)

Service procedures for other float-type carburetors are not significantly different. You should encounter no problems with the Carter Model N (parts are still available from Kohler), the several Tillotson patterns, and the Pincor.

CARBURETORS WITH INTEGRAL FUEL PUMPS

Briggs & Stratton builds three models of the *Pulsa-Jet* carburetor. All function on the same broad principle, but there is little part commonality between them. You will find these carburetors on late-production rotary and riding mowers. We will direct our discussion to the most popular model.

Pulsa-Jet carburetors feature a diaphragm operated by manifold vacuum that lifts fuel out of the tank and discharges it into a reservoir. (See Fig. 5-30.) There are two fuel pickup tubes, one for the pump section and one for the carburetor proper.

The reservoir keeps the fuel level constant, regardless of the level in the tank. This standpipe system is reminiscent of the earliest carburetor designs.

Diaphragm flaps on this unit double as check valves.

Fuel pickup tubes are threaded and can be removed with the appropriate wrench. No check valves are employed. Check the O-rings at the carburetor bore and crankcase breather connection. Leaks at either point will cause lean running. Replace the diaphragm with each overhaul.

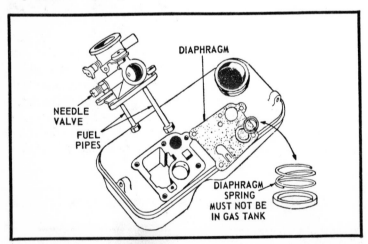

Fig. 5-30 "Pulsa-Jet" showing reservoir in the top of the tank (Courtesy Briggs & Stratton)

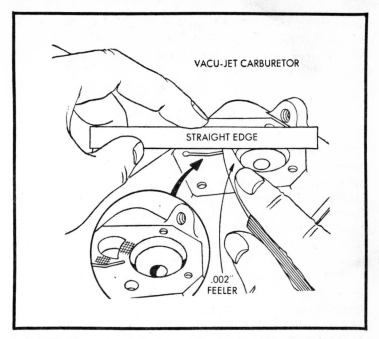

Fig. 5-31 Check the shaded areas for trueness (Courtesy Briggs & Stratton)

Note: If the tank is dry or if the machine has been tilted, a half-dozen pulls on the cord will be required for starting.

Initial adjustment is $1^1\!/_2$ turns open; the idle should be set to 1750 rpm.

All 92000 series B&S engines built since August 1968 feature an automatic choke. This choke is fitted to Vacu-Jet and Pulsa-Jet carburetors, but is more common on the latter. The choke is normally held closed by a spring. When the engine catches, manifold vacuum overcomes the tension of the spring and forces the choke open. You can verify its operation by removing the air filter and replacing the stud (the valve should be closed). Crank vigorously—the choke disc should flutter with each intake stroke.

Checkpoints are:

1) The air cleaner stud. If it is bent, it may lock the disc open or closed.
2) The choke shaft (for sticking).
3) The choke spring, to see if it's too short (Pulsa-Jet spring: $1^1\!/_8$ to $1^7\!/_{32}$ mx; Vacu-Jet: $^{15}\!/_{16}$ to 1 in.)

4) The fuel and oil; accumulations in the vacuum chamber may be caused by a leaking diaphragm or by irregularities in the machined surface on the tank.

5) Preload.

6) The disc-type blade. If it is heavy and keeps spinning after shutdown, it may flood the engine in the process. The solution (partial, at least) is to adjust the main jet ⅛ turn leaner than best mixture.

The shaded areas of Fig. 5-31 represent critical areas. Low spots will allow fuel to be drawn into the vacuum chamber and dampen the choke action. Discard the tank if you can insert a 0.002 in. feeler gage under a machinist's straightedge at either of these points. The choke link fits into the diaphragm and is secured at the bottom by a small metal tab sandwiched between the spring and the diaphragm. The spring fits into a recess (Fig. 5-32) in the tank. Guide the link up to the carburetor and connect it to the choke shaft. Run the screws connecting the carburetor to the tank down about 2 turns. Then, insert a ⅜ in. bolt between the choke disc and the air horn as shown in Fig. 5-33. (You will have to remove the air horn gasket for this operation.) Tighten the screws evenly, in an alternating crisscross fashion. The purpose of the bolt is to

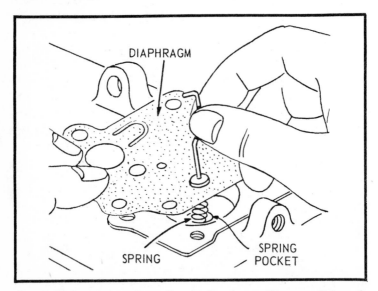

Fig. 5-32 Positioning the spring in the recess (Courtesy Briggs & Stratton)

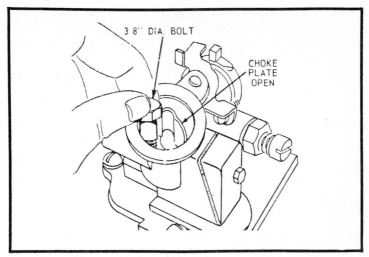

Fig. 5-33 Preloading the choke diaphragm (Courtesy Briggs & Stratton)

preload the diaphragm against the spring. Remove the bolt, install the assembly on the engine, and check the choke action.

A few Tecumseh carburetors have integral fuel pumps, in several variations. These pumps respond to crankcase pressure pulses and deliver fuel in direct porportion to engine speed. One type, illustrated in Fig. 5-5, employs a balloon-like element. On the downstroke, the crankcase is pressurized. Some of this pressure is diverted through a port in the carburetor mounting flange and inflates the "balloon." The inlet check valve closes—you can see it at the upper right of Fig. 5-34—and the outlet opens. Fuel leaves the diaphragm chamber and enters the float chamber. On the upstroke the pump element deflates: the inlet check valve opens and the outlet closes. Fuel is drawn into the pump chamber; it will be expelled on the downstroke.

To check the operation of these pumps, first determine if the impulse port is clear and aligned properly with the pulse port at the carburetor flange. Next check the condition of the pump element. If it looks doubtful, replace it. You can determine if the valves are leaking by alternately sucking and blowing through them. Use a length of fuel line cut flat, and placed hard against the impulse port on the flange. Place a finger over, and closing, the fuel inlet fitting. Suck the air out through the fuel line; the inlet valve should close. The carburetor body valve is tested at the inlet fitting. Install the fuel line on the fitting and draw gently. The valve should close and hold pressure. If it does not, blow to unseat the valve and repeat the test. Replacement valves are available, in case buying new ones is your only option left. These carburetors should not be cleaned chemically unless you intend to install new valves.

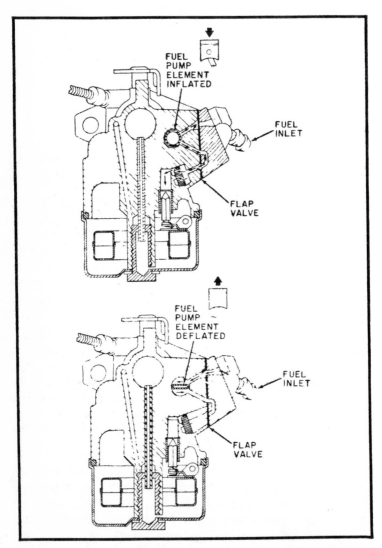

Fig. 5-34 (A) Pump inflated, piston on the downstroke. (B) Pump deflated, piston on the upstroke (Courtesy Tecumseh Products Co.)

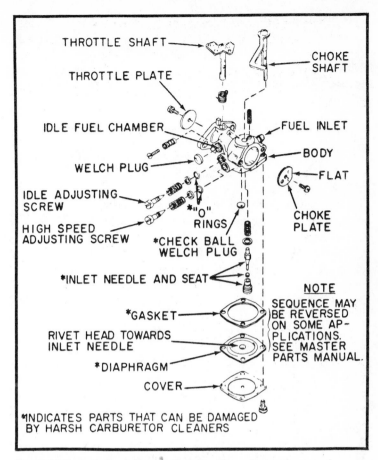

THROTTLE SHAFT

CHOKE SHAFT

THROTTLE PLATE

IDLE FUEL CHAMBER

FUEL INLET

BODY

WELCH PLUG

FLAT

IDLE ADJUSTING SCREW

HIGH SPEED ADJUSTING SCREW

*"O" RINGS

*CHECK BALL WELCH PLUG

CHOKE PLATE

*INLET NEEDLE AND SEAT

NOTE

SEQUENCE MAY BE REVERSED ON SOME AP-PLICATIONS. SEE MASTER PARTS MANUAL.

*GASKET

RIVET HEAD TOWARDS INLET NEEDLE

*DIAPHRAGM

COVER

*INDICATES PARTS THAT CAN BE DAMAGED BY HARSH CARBURETOR CLEANERS

Fig. 5-35 Typical diaphragm carburetor. These devices have undergone numerous modifications. Models with a primer have a check valve inside of, or behind, the fuel inlet fitting. Others have a check ball in the main nozzle (similar to the check ball in pickup tubes on suction-lift designs). The ball may or may not be serviceable, depending upon model. (Courtesy Tecumseh Products Co.)

Diaphragm-Type Carburetors

Tecumseh, Tillotson, Walbro, and Bendix have all built diaphragm carburetors for small engines. A number of variations have been used on lawn equipment during the last decade. Some have an integral fuel pump in the form of a second diaphragm, but most were gravity-fed.

The diaphragm is open to the atmosphere on one side and subject to crankcase pulses on the other. As the engine runs, the diaphragm moves up and down, unseating the inlet needle. Metering, cold-start, and other systems are entirely conventional.

The Tecumseh design is illustrated in an exploded view in Fig. 5-35. The diaphragm is vented through a central hole in the cover. The carburetor can be flooded by inserting a length of wire through the vent and gently pushing it upward. The best procedure is to remove the unit from the engine before any major service operations are undertaken. However, the diaphragm can be changed in-place by loosening the four cover screws. You will have to use an offset screwdriver on rotary lawnmowers because of close proximity to the deck. Install the new diaphragm *per the instructions in the replacement package.*

Care should be taken with the inlet needle—it's very delicate. Remove the seat with a $^9/_{32}$ in. six-point socket. (You may have to grind the socket sides slightly.) Treat the needle spring gingerly, being careful not to compress or bend it. The *idle* and *high-speed* screws should not be interchanged, even though the threads are identical. Study the jets under a strong light; it is not unusual to find a needle tip snapped off in the jet orifice. Examine the welch plugs, and seal with epoxy any that are loose. Initial adjustment is ¾ to one turn out on both screws. Figure 5-36 illustrates a Tillotson twin-diaphragm carburetor.

EXTERNAL FUEL PUMPS

Briggs employs a mechanical pump on a number of their larger engines. The pump is driven by a cam located on the crankshaft (Fig. 5-37). To test this or other lawnmower pumps, ground the ignition securely, disconnect the discharge line, and crank the engine. Manufacturers do not supply flow and pressure data, but anything more than a dribble is sufficient.

The only routine service required is the replacement of the diaphragm. To do this, remove the pump from the engine, and take out the six screws holding the diaphragm cover; then drive out the fulcrum pin (in either direction), and unhook the pump lever from the diaphragm. Install the new diaphragm over the spring, and lock it into the pump lever. Replace the pin and assemble. These pumps are a bit tricky to mount; the arm must ride in the grooves on the crankshaft as illustrated, and it might take several tries to get it right.

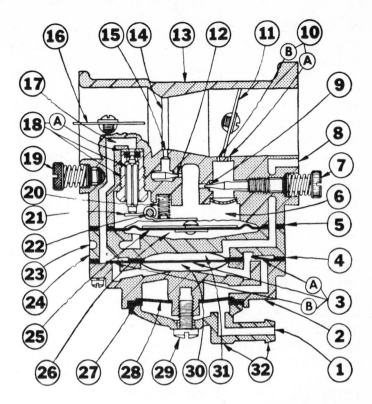

1	Hose nipple	10A	Idling outlet passage	20	Inlet compression spring
2	Fuel pump body	10B	By-pass passage	21	Inlet control lever
3	Fuel pump diaphragm	11	Butterfly valve	22	Pivot pin
3A	Inlet valve on pump diaphragm	12	Main jet orifice	23	Vent hole
		13	Carburettor housing	24	Diaphragm housing
3B	Outlet valve on pump diaphragm	14	Throat of carburettor (venturi)	25	Diaphragm
		15	Main fuel outlet	26	Air chamber
4	Fuel pump gasket	16	Choke	27	Filter gasket
5	Diaphragm housing gasket	17	Fuel passage from pump to control chamber	28	Fuel strainer
6	Control chamber			29	Filter housing screw
7	Adjusting screw for idling jet	18	Inlet needle and seat	30	Fuel chamber
8	Impulse passage	18A	Copper seal	31	Pulsation chamber
9	Idling jet orifice	19	Adjusting screw for main jet	32	Fuel filter housing

Fig. 5-36 Tillotson, with integral fuel pump. The inlet control lever (21) should be parallel to the diaphragm chamber as shown by the arrows in the inset. Adjust by bending on the inlet needle side. Do not apply force to the needle.

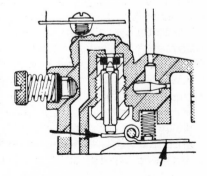

Clinton and Tecumseh have opted for an impulse pump. The Clinton is usually mounted directly on the block, above the oil level. Tecumseh drives theirs through an impulse line. Both have check valves in the form of flaps cut into the diaphragm.

AIR FILTERS

Rotary lawnmowers and edgers usually have to work in a cloud of dust. It is vital that this dust be kept out of the engine. You can appreciate the magnitude of the filtering necessary when you consider that for every gallon of gasoline consumed, the engine inhales 10,910 cubic feet of air (at a 15:1 mixture ratio).

The filter doubles as a silencer and a firescreen. The latter function is highly important since all engines will backfire occasionally. In 2-cycle applications, the filter helps keep the deck and engine castings tidy by containing the fuel/oil mist which hovers around the carburetor throat.

Early filters were quite primitive, both in terms of filtration efficiency and pressure drop. When working properly, these filters tended to strangle the engine. Some were made of composition board molded into a cylinder and sealed at the end. Others were cakes of compressed aluminum shavings.

The oil-bath filter is a welcome improvement. Figure 5-37 shows a Kohler design. The element—a fibrous mesh—is wetted by an oil supply carried at the base of the filter. Most foreign particles are trapped in the mesh. However, some filtration occurs as the incoming mass changes direction. Because dust particles have mass, they continue in a straight line and fall into the oil.

Depending upon operating conditions, these filters should be serviced every 2—8 hours of use. Clean the element and the filter body in solvent, air-dry, and add oil to the reservoir. Be careful not to overfill, as the excess will be drawn into the carburetor. Carefully inspect the mounting gaskets and replace as needed. A bent mounting stud will hasten gasket wear.

Oil bath filters are fast being displaced by the new types. Pleated paper filters are claimed to be able to stop particles as small as 25 microns (0.025 mm) in diameter. Polyurethane filters are as efficient, but not as predictable. Filtration depends upon how much oil is used to wet the element. Too

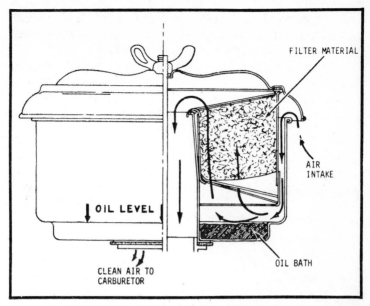

FILTER MATERIAL

AIR INTAKE

OIL LEVEL

OIL BATH

CLEAN AIR TO CARBURETOR

Fig. 5-37 Oil bath filter. The bends in the inlet tract centrifuge the dust particles out of the air stream (Courtesy Kohler of Kohler)

much oil brings about excessive pressure drops and rich mixtures.

Figure 5-38 demonstrates the recommended cleaning and oiling procedure for Briggs engines. Wash the element in solvent, or in detergent and water. Wrap the element in a cloth and squeeze it dry (Fig. 5-38B). Saturate with engine lube (C),

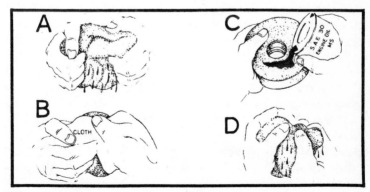

Fig. 5-38 Recommended polyurethane filter cleaning procedure (Courtesy Briggs & Stratton)

and dry by squeezing (D). Clinton employs a similar polyurethane foam filter but does not suggest that the element be wetted with oil—kerosene alone is enough.

Paper filters *should not be wetted under any condition.* Water or oil will cause the fibers to swell and block the tiny air passages. For this reason, paper filters are rarely specified for 2-cycle engines, although they can be used if the air horn has an oil baffle. Certain Hahn-Eclipse riding mowers had a special muffler which allowed exhaust gases to be introduced back into the carburetor. If you encounter any of these machines in the field, substitute a polyurethane filter for the paper element which was originally supplied. It has been found that the exhaust contains enough oil to damage paper elements.

Tecumseh offers a precleaner kit for extremely dusty conditions which can be used with either type of element. The better Craftsman mowers have this feature (normally an option) as standard equipment. Another variant is the Kleen-Aire system illustrated in Fig. 5-39.

Caked dirt can be removed by rapping the element sharply on a hard surface. An air gun (regulated to 30 psi) is also helpful. Replace the filter when more than 20% of the surface is oil-stained; if the filter is cracked or torn or refuses to pass sunlight, substitution is in order.

GOVERNORS

Lawnmower engines are constant-speed devices, designed to operate between 1750 rpm at idle and 3400 rpm at full throttle without load. Larger engines often have a lower no load speed. In any event, blade tip velocity must not exceed 19,000 feet per minute. The governed no-load, full-throttle speed should be periodically checked with an accurate tachometer against factory specifications.

Two types of tachs are available. Electronic tachs for single-cylinder engines are quite expensive and, in some applications, may require removing the flywheel to attach the leads. Most shops use a vibrating-reed tach, which is held in the exhaust blast. The reed is adjustable in length, and vibrates in sympathy with power pulses. These tachs are fairly accurate, but you should allow yourself a 200 rpm fudge factor.

The following rpm table has been computed by Briggs & Stratton.

BLADE LENGTH. in.	MAXIMUM SPEED. rpm
18	4032
19	3820
20	3629
21	3456
22	3299
23	3155
24	3024
25	2903
26	2791

Governors fall into two broad categories depending upon the speed sensor mechanism used. Most engines still employ an air vane governor similar in operation to the one in Fig. 5-40. The governor spring tends to pull the throttle plate open. As engine speed increases, the flow of air from the flywheel

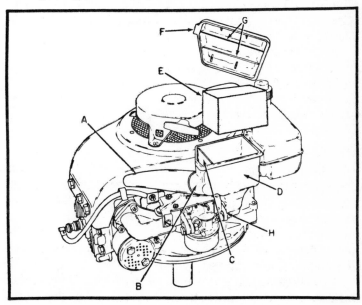

Fig. 5-39 The patented "Kleen-Aire" system. Air is picked up under the cowling, in reverse of blower flow. Entering air velocity is rated low since the blower and the snorkel work in opposition. Air velocity is further slowed by a baffle at C. Dirt particles drop out and are collected in the pan area at B. The polyurethane element (E) is wetted with two tablespoons of oil. Channels at G open the whole upper surface of the filter to the air stream (Courtesy Tecumseh Products Co.)

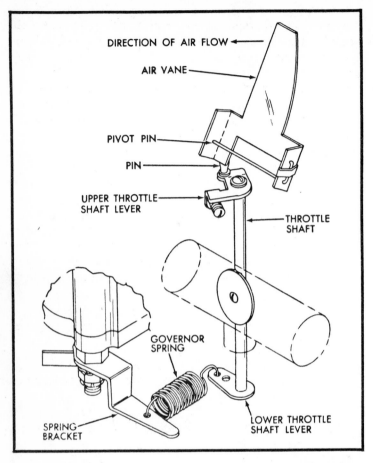

Fig. 5-40 Air-vane governor operation (Courtesy Tecumseh Products Co.)

fan also increases, and tilts the vane to the right. The throttle plate is in balance between the spring pulling it open and the vane pushing it closed. The operator can override the governor with the manual throttle control connected to the spring bracket. As the bracket swivels it changes spring tension. In this particular design, maximum no-load speed is determined by spring tension and by the leverage the spring exerts on the lower throttle shaft lever. The second attachment point shown in the drawing provides an option.

Other air vane governors are more complex, and may have two or more springs and several wire links between the vane and the carburetor. When troubleshooting, first check the vane for freedom of movement. Most employ sheet-metal tabs for hinges and can easily jam. (Be careful not to cut your fingers on the vane.) Next, examine the linkages for straightness. Some are deliberately bent in a single plane—but the link should not look like a length of tired spaghetti. Replace the spring if it shows signs of fatigue or coil separation or if it has battered ends. Spring removal can be tricky and requires the same two-dimensional twist technique used in the "bent-nails" puzzle (Fig. 5-41). Do not force the spring.

Air vane governors are tardy in response and, unless made frightfully complex with compensating links and springs, have a tendency to surge. While all manufacturers employ centrifugal governors on their larger 5 hp-and-up engines, Tecumseh and OMC were the first to extend the principle to lawnmower and edger powerplants.

The Tecumseh 4-cycle governor uses paired centrifugal weights in the crankcase. As engine speed increases, the weights move outward and, acting through a lever and link, tend to close the throttle (Fig. 5-42). Adjustment for vertical shaft engines is as follows:

1) Loosen the pinch bolt.
2) Turn the governor shaft to the left (counterclockwise) with pliers. *Do not force.*
3) Move the governor on the shaft. Both parts should be held full left. The connecting link should force the throttle plate fully open.
4) Tighten the pinch nut and test the engine with a tachometer.

The same sequence applies to horizontal-crankshaft engines, except that the movement is to the right. The rationale is that the governor shaft must be seated against the weights and the throttle plate be fully open.

Early Power Products and OMC series-A engines are similar in that the centrifugal weights are mounted below the case on the power-takeoff end of the shaft. Figure 5-43 shows the arrangement for both engines. Maximum free-running speed should be kept on the conservative side of 3200 rpm. Loosen the setscrew and move the fixed collar relative to the shaft. Sliding it upward increases the speed; downward lowers it.

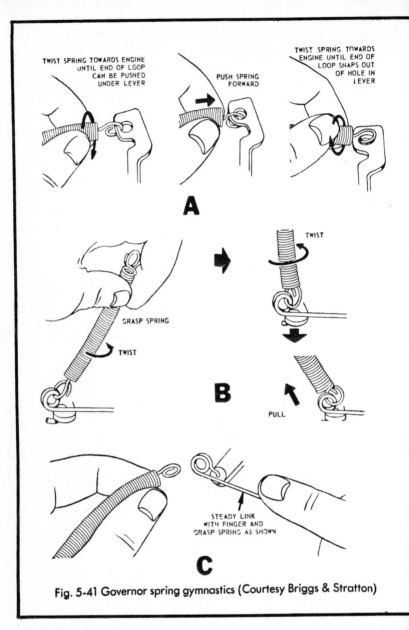

TWIST SPRING TOWARDS ENGINE UNTIL END OF LOOP CAN BE PUSHED UNDER LEVER

PUSH SPRING FORWARD

TWIST SPRING TOWARDS ENGINE UNTIL END OF LOOP SNAPS OUT OF HOLE IN LEVER

A

GRASP SPRING

TWIST

TWIST

PULL

B

STEADY LINK WITH FINGER AND GRASP SPRING AS SHOWN

C

Fig. 5-41 Governor spring gymnastics (Courtesy Briggs & Stratton)

These governors are very responsive; engine speed can be expected to remain almost constant regardless of load. If the engine bogs down or overspeeds, check the linkage for freedom of movement. The OMC design has a dust seal which

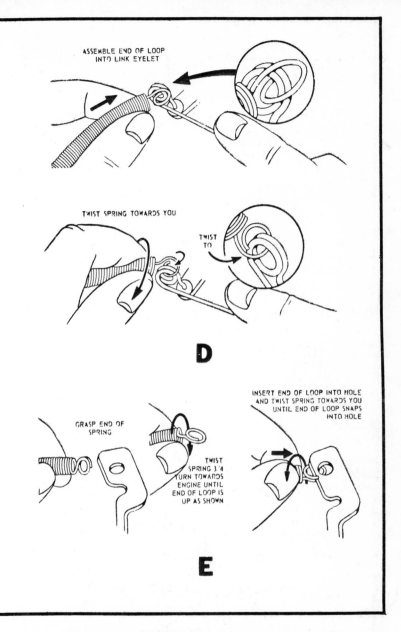

ASSEMBLE END OF LOOP
INTO LINK EYELET

TWIST SPRING TOWARDS YOU

TWIST
TO

D

GRASP END OF
SPRING

TWIST
SPRING 3/4
TURN TOWARDS
ENGINE UNTIL
END OF LOOP IS
UP AS SHOWN

INSERT END OF LOOP INTO HOLE
AND TWIST SPRING TOWARDS YOU
UNTIL END OF LOOP SNAPS
INTO HOLE

E

can deteriorate and bind the throttle lever. It may be necessary to pull the cover and clean the scissor linkage.

Both manufacturers have moved the governor mechanism to a spot under the flywheel on current production models

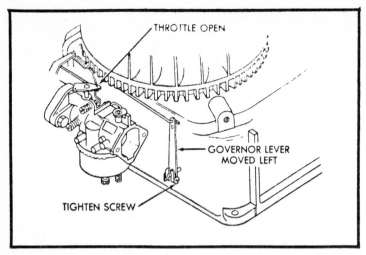

Fig. 5-42 Governor adjustment for Lauson engines (Courtesy Tecumseh Products Co.)

(Fig. 5-44). This leaves room for a drive pulley on the OMC design, and enables the basic Power Products block to be adapted to numerous roles, including chainsaw and outboard applications.

The Power Products governor is illustrated in the cutaway view of Fig. 5-45. Gross adjustments are made at the mounting bracket screw with the engine *stopped*. Once the screw is loosened, the assembly can be moved up and down. Moving it toward the flywheel increases engine speed. Make fine adjustments by bending the rod. Operate the engine as slowly as possible consistent with good grass cutting.

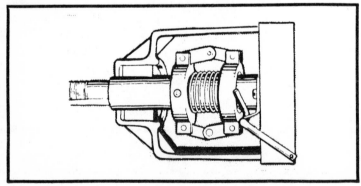

Fig. 5-43 Lawnboy Series A governor (similar to early Power Products) (Courtesy OMC)

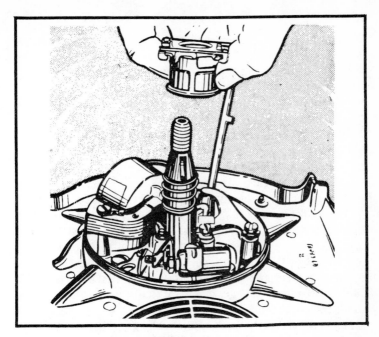

Fig. 5-44 Late-model Lawnboy governor. The spring around the crankshaft stub controls engine speed. The 2800 rpm spring is color-coded; the 3200 spring is uncolored (Courtesy OMC)

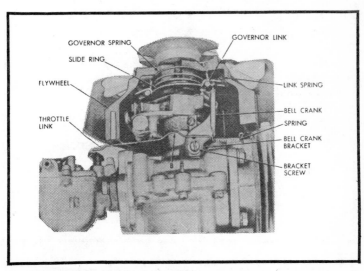

Fig. 5-45 Power Products flywheel governor (Courtesy Tecumseh Products Co.)

Some of these engines do not have a governor as such. Instead, they are fitted with an ignition cutout which limits free-running speed but does not compensate for loads. A slip ring shorts the primary circuit. Maximum speed is a function of engine design and application. Check against the appropriate factory specs.

The OMC series-D speed control mechanism is somewhat complex. The operator can vary the no-load speed between 2500 and 3200 rpm by means of a knob located below the gas tank. Primary settings are made with the flywheel removed.

Larger Clinton engines employ a centrifugal governor driven by the camshaft. Figure 5-46 shows the mechanism in the throttle-closed position. Load has suddenly been removed and the throttle plate snaps shut in response to the outward movement of the spinning weights. Clinton governs all such engines to 3600 rpm maximum. An ungoverned engine will hit 7000 before the rod lets go. It is vital that the correct spring be used, that it is connected properly, and that the adjustment screw is set correctly. Look closely at Fig. 5-48. If the $^1/_{32}$ to $^1/_{16}$ in. clearance between the throttle plate and carburetor body is not maintained, damage is possible to the flyweights. Adjust

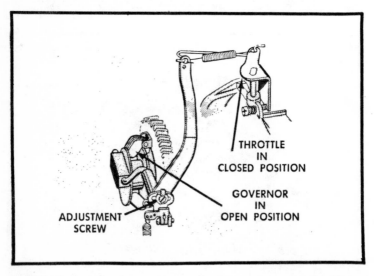

THROTTLE
IN
CLOSED POSITION

GOVERNOR
IN
OPEN POSITION

ADJUSTMENT
SCREW

Fig. 5-46 Centrifugal governor mechanism used on certain Clinton engines, but similar in concept to those built by other manufacturers (Courtesy Clinton Engines Corp.)

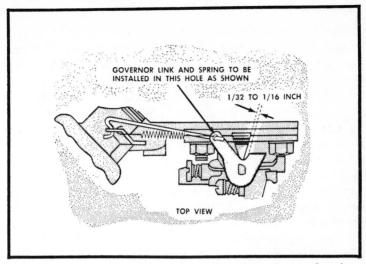

Fig. 5-47 Throttle lever clearance (Courtesy Clinton Engines Corp.)

the screw with the engine shut down, and adjust tension on the governor spring to this dimension. Then make a final check with a tach.

Briggs engines are commonly equipped with air vane governors, but you will find the more sophisticated centrifugal types on the larger blocks. Figure 5-49 locates the adjustment points for current vertical- and horizontal-shaft models. *Do not run these engines with the links disconnected*; serious damage to the weights could result.

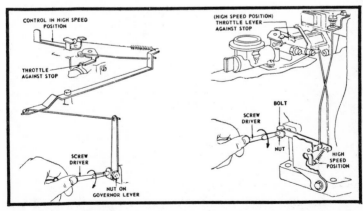

Fig. 5-48 Adjustment drill for Briggs' centrifugal governors (Courtesy Briggs & Stratton)

FUEL TANKS AND FITTINGS

Steel tanks are almost a thing of the past, at least so far as light mowers are concerned. Few mechanics regret their passing, since the steel used—even when coated with zinc—invariably fell prey to water (present in some degree in all gasoline supplies). Rust particles collected at the bottom of the tank to be swept out into the fuel line and carburetor. Eventually the rust would bite deeply enough into the tank to cause leaks.

As a rule, metallic tanks cannot be repaired. The accumulation of debris can be removed with cleaner, but it quickly develops again. Briggs & Stratton tanks have always been crimped and sealed with filler. These tanks will not tolerate long steeping in the chemical bath. Leaks can sometimes be stopped with an application of epoxy to both top and bottom surfaces. However, the repair cannot be guaranteed. **Do not attempt to mend a leaking tank with solder**. This is so dangerous that it can only be described as stupid. It is impossible to remove all standing fumes from a gasoline tank. Steam, water flush, CO_2, or any other expedient is never 100% certain. Should the tank ignite, it will explode with tremendous force, and probably make the last sound you'll ever hear.

Plastic tanks should be periodically checked for leaks and mounting strap tension. Aside from that, no repairs are possible.

Chapter 6

Manual Starters

Electric motors and *Rankine cycle* steam engines develop maximum torque at zero rpm. In contrast, internal combustion engines develop no torque at all at zero rpm. Some energy must be put into the engine to initiate the *Otto working cycle*. Magneto-sparked lawnmower engines must be spun to about 90 rpm before fuel, compression, and spark become self-supporting.

Rewind Starters

Vintage engines had a knotched pulley to accept the knotted end of a rope. The beauty of this system was its utter simplicity. The only breakable part was the rope, and that was easy enough to replace.

The next step, pioneered by Jacobson in 1928, was to spring-load the cord so it automatically rewound itself over the pulley. Some heavy engineering was required for this seemingly simple design change. The starter housing is mounted on the shroud; all starting loads are absorbed by engine sheet metal. In some cases, the shroud had to be redesigned to accept these loads. The starter mechanism consists of a flat coiled spring anchored at one end to the housing. The other end is secured to the pulley so that the spring winds as the pulley rotates. During the first few degrees of rotation a clutch is engaged to lock the pulley and crankshaft together. When the engine fires, the clutch overruns.

Rewind starters can be mounted in line with the shaft or at right angles to it. The latter method has certain advantages for vertical-shaft engines; starting is easier—one merely bends from the waist and pulls upward on the cord. With the exception of Jacobson, these starters engage the flywheel by means of a reduction gear. Torque is multiplied at a tradeoff in rpm. The Jacobson design employs a starter pulley directly over the flywheel hub with the rope routed upward to give vertical action.

Whatever rewind starter your mower has, certain general troubleshooting procedures hold:

1) Noise from the starter, as the engine runs, means a dry bearing (Briggs) or misalignment between the starter and the flywheel hub.

2) Stiff action indicates lack of lubrication. The spring and internal parts should be greased *lightly*; too much lubricant will soak the rope, collecting dirt and binding the action.

3) Refusal to engage or slippage is evidence of clutch problems. It won't be necessary, however, to disassemble the whole starter to relieve these symptoms.

4) Reasons for failure to disengage are academic, since the starter will have exploded. Suspect clutch or alignment difficulties.

5) Refusal to rewind is indicative of a weak main spring, insufficient preload, or starter housing misalignment. A contributory cause can be an oil-sodden rope.

6) Chronic problems with a particular starter are usually traceable to the user—not the design. The cord should be pulled out smoothly and held while it retracts. Otherwise, expect cord and spring failures.

7) Finally, work carefully. The spring can give you a nasty thrashing if it escapes its housing.

Briggs & Stratton

Briggs starters are inhouse designed and are significantly different than those used by other manufacturers. The starter is integral with the blower housing. A slot in the housing anchors the rewind spring (Fig. 6-1).

The clutch housing (Fig. 6-2) threads over the crankshaft and secures the flywheel. The ratchet is free to turn; it rides on a bushing on the crankshaft stub and its outer end mates in a socket with the pulley. When the pulley is rotated, the ratchet turns and traps a ball bearing against the side of the housing, so that both parts—ratchet and clutch housing—turn together. When the engine fires, the housing moves relative to the ratchet, and the bearing disengages. As homely as this starter looks in the metal, it is conceptually elegant.

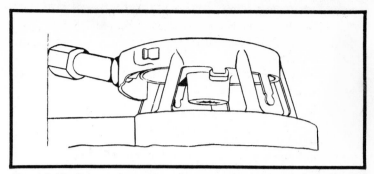

Fig. 6-1 The Briggs starter. Note the exposed spring end which dictates special repair procedures (Courtesy Briggs & Stratton)

To dissassemble, remove the blower housing and place it upside down on the bench. With a pair of pliers, pull the anchored end of the spring as far out of the housing as it will come (Fig. 6-3), or untie the handle and allow the cord to retract in a controlled fashion. This is an important step, analogous to unloading a gun before cleaning it. All rewind and impulse starters are under spring tension. Insofar as possible, this tension should be released prior to disassembly. Wise technicians further protect themselves by wearing safety glasses; I recommend you follow their example.

The pulley is located on the housing by several tangs (in the tradition of Marx toys). Straighten the tangs enough to lift the pulley. If you break a tang, notice that Briggs has

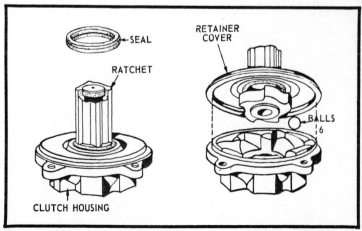

Fig. 6-2 Clutch assembly which doubles as a flywheel nut (Courtesy Briggs & Stratton).

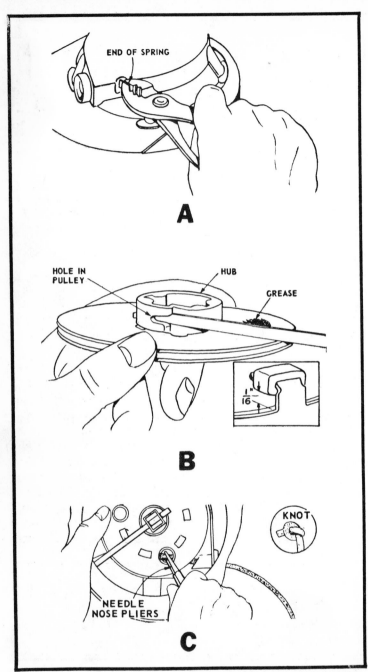

END OF SPRING

A

HOLE IN PULLEY

HUB

GREASE

$\frac{1"}{16}$

B

KNOT

NEEDLE NOSE PLIERS

C

Fig. 6-3 Service operations—see text (Courtesy Briggs & Stratton)

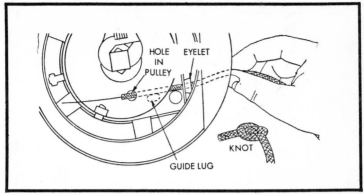

Fig. 6-4 Inserting the rope. The dotted lines represent a guide bar which was cast into the older pulleys (Courtesy Briggs & Stratton)

thoughtfully provided spares. The spring can now be unhooked from the underside of the pulley.

The tricky aspect of Briggs starters is the assembly. Secure the blower housing to the bench by nails, or mount it in a big vise. Thread the spring through the hole in the housing and connect it to the pulley. Cut a piece of one-by-one to fit the pulley socket. Wind the pulley 13¼ turns counterclockwise.

It helps to bend the tangs low so they almost rub the pulley. Vertical slack will allow the spring to wind crosswise. As the end approaches, turn the pulley very slowly to give the spring an opportunity to snap into the narrow portion of the slot. While holding tension on the pulley, thread the rope through the housing grommet and into the pulley with your third hand. Old-style pulleys had an internal guide lug—shown by dotted lines in Fig. 6-4. The rope must be snaked behind this lug. A probe made of piano wire is useful here. (See Fig. 6-5.)

Melt the end of the cord with acetone or an open flame, and knot as shown. Secure the handle and release the pulley, braking it with your thumbs. If you have done everything

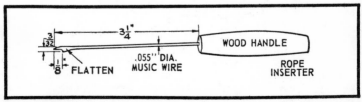

Fig. 6-5 A rope tool which is useful for all starter makes (Courtesy Briggs & Stratton)

correctly, the rope will wind itself around the pulley. Open the tangs to provide a running clearance of $1/16$ inch between the nylon bumpers and the pulley face.

The Briggs vertical-pull starter is disarmed by lifting the rope out of the pulley groove and turning the pulley counterclockwise 2 or 3 revolutions (Fig. 6-6). Pry the plastic cover free and detach the rope anchor. Do not engage the starter with the cover off and the anchor in place.

Remove the rope guide (held by a self-tapping screw) and note the position of the link before further disassembly. Check the link—it is the most vulnerable part of this starter. It should move the gear to both extremes of travel. If you have any doubts about the link, replace it. Don't waste time trying to straighten it.

Assemble the rope into the pulley, using the method demonstrated in Fig. 6-7. Space is limited, and the end of the line must not extend more than $3/16$ inch past the knot. Install the spring (Fig. 6-8), wind counterclockwise until the rope is fully retrieved (Fig. 6-9), and tighten the anchor nut. Snap the cover in place and preload the mechanism by slipping the rope

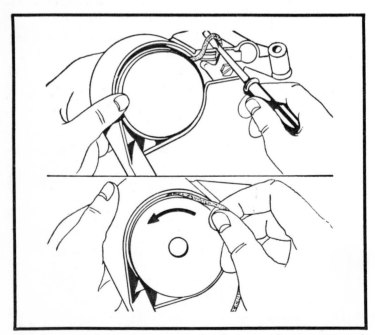

Fig. 6-6 Disarming the Briggs vertical-pull starter (Courtesy Briggs & Stratton)

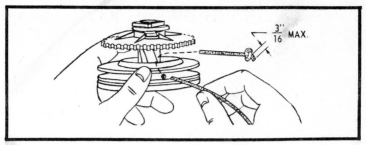

Fig. 6-7 Starting the rope into the pulley. Note that only ³/₁₆ in. is allowed on the tag end (Courtesy Briggs & Stratton)

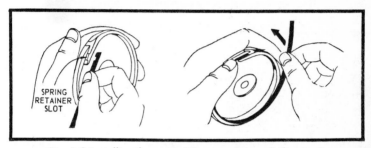

Fig. 6-8 Installing the spring (Courtesy Briggs & Stratton).

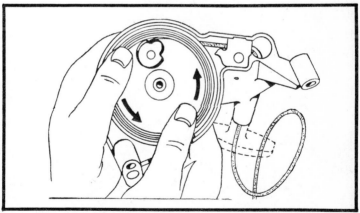

Fig. 6-9 Initial spring wind (Courtesy Briggs & Stratton)

out of the pulley and winding the pulley 2 or 3 turns clockwise. Figure 6-10 illustrates the process.

Fairbanks-Morse

Fairbanks-Morse starters have been used on a variety of engines, including Kohler, Tecumseh, and West Bend. Their

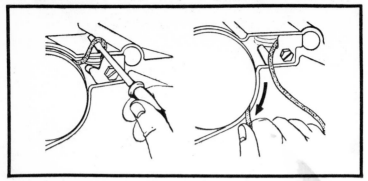

Fig. 6-10 Final spring wind for preload (Courtesy Briggs & Stratton)

distinguishing feature is the clutch assembly with its two friction shoes (No. 12 in Fig. 6-11). Most starter clutches employ some sort of positive engagement against a bearing (as we have seen in the Briggs) or against teeth milled internally on the crankshaft hub. The Fairbanks-Morse pattern engages by friction alone. The shoe ends are sharpened and work against a smooth-walled hub. Another

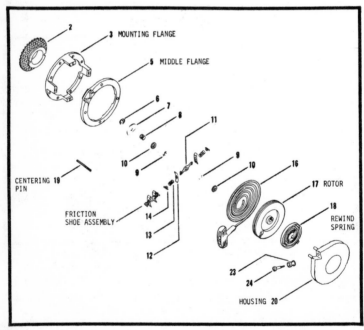

Fig. 6-11 Fairbanks-Morse starter in exploded view (Courtesy Kohler of Kohler)

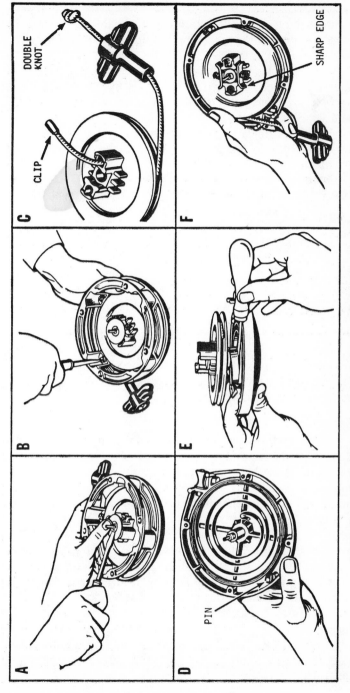

Fig. 6-12 Fairbanks-Morse starter service operations (Courtesy Kohler of Kohler)

distinguishing mark is the stainless-steel cable, found on some models instead of the nylon line. Replace the cable at the first evidence of fray or wear.

Some F-M starters, including the one depicted in Fig. 6-11, have a small coil spring (part 8), located behind the thrust washer (7). When you remove the Tru-Arc clip (6), be sure to hold your thumb over the thrust washer, as is being done in Fig. 6-12A. Release tension on the main spring by disconnecting the handle and allowing the pulley to wind, braking it with your thumbs.

Note the position of the friction shoes. They *are not* assembled as shown in Fig. 6-11. Follow the drawing in Fig. 6-13. If the faces have dulled, sharpen lightly on a grinder. Remove any burrs.

The flanges (5 and 3) are removed as shown in Fig. 6-12B. Ropes for larger engines are fitted with a steel clip. Smaller models employ the familiar knot, with the tag ends melted. Uncoil the spring one winding at a time. Replacement springs are prewound on a heavy steel form to make installation easier—and safer. Position the form and spring over the pin; cut the masking tape which holds the spring in the form, and push down evenly on all sides. Note the lay of the spring. It is disconcerting to install a spring backwards.

Preload varies with the starter model, but should not be more than five turns in any case.

Eaton

Eaton-pattern starters are used on many Tecumseh and Kohler products. The originality of the design is in the engagement mechanism. A brake spring causes the retainer cup to turn and cam out one or more pawls which engage serrations in the flywheel hub. Figure 6-13 is an exploded view of the starter most frequently encountered. The brake spring is the star-shaped object numbered 7. The cup is 5, and the pawl (or dog) is 8. Larger engines, which require more torque input, have 3-dog starters, but otherwise function as the one shown. Figure 6-14 illustrates a light-duty starter found on small Lauson and Power Products blocks. Figure 6-15 represents a modification of the basic pattern. The brake spring (5) is a coil and the spring-and-keeper (9) is to be considered as a single assembly. *Do not remove the spring from the keeper.* This starter features a centering pin which runs on a nylon bearing (2).

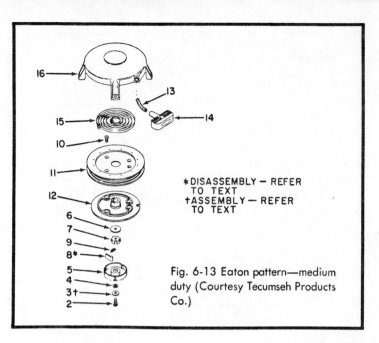

*DISASSEMBLY — REFER
 TO TEXT
†ASSEMBLY — REFER
 TO TEXT

Fig. 6-13 Eaton pattern—medium duty (Courtesy Tecumseh Products Co.)

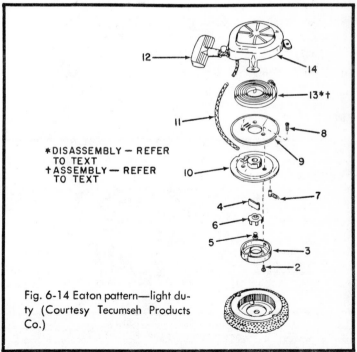

*DISASSEMBLY — REFER
 TO TEXT
†ASSEMBLY — REFER
 TO TEXT

Fig. 6-14 Eaton pattern—light duty (Courtesy Tecumseh Products Co.)

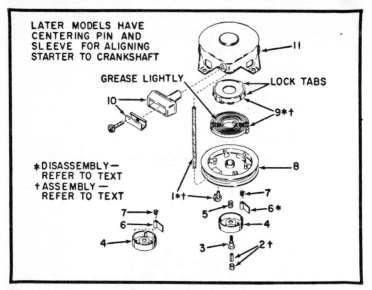

Fig. 6-15 Eaton pattern—unitized main spring. Do not attempt to separate the spring and holder (NO. 9) (Courtesy Tecumseh Products Co.)

Some parts on the three units shown, and other Eaton starters, are interchangeable; but you should purchase replacements by engine type and model number. Logistics is much more complicated than with inhouse designs.

Remove the unit from the engine and release spring tension by prying the rope up and over the notch in the pulley (Fig. 6-16). Loosen the retainer screw at the bottom of the unit.

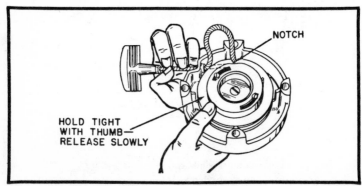

Fig. 6-16 Disarming the Eaton. This technique may be used with almost all rewind starters, regardless of make (Courtesy Tecumseh Products Co.)

If the screw was already loose (i.e., torqued to less than 45–55 ft-lb) you have found at least part of the problem. A loose retainer will not cam the dog into engagement. Lift the retainer cup and note the position of the dog(s). The curled end points upward and the working edge points out. Inspect the dog-return spring (9). Sometimes these springs break. (Be gentle—the spring can slip out of its housing.)

Observe the lay of the spring; most of these starters can be used in right- or left-hand rotation engines. New springs are packaged with retainer rings, although it is possible to rewind a spring a coil at a time. At the risk of redundancy, allow me to remind you again that the main spring shown in Fig. 6-15 is part of an assembly. It should not be rewound or otherwise separated from its housing.

To assemble, follow the numbering on the three illustrations in reverse order. Preload is approximately 6 turns (the exact tension varies somewhat as the spring weakens with age). The centering pin and bushing (2 in Fig. 6-15) are placed in the crankshaft center. The bushing should be installed to a depth of ⅛ inch over the pin. Start two of the mounting screws 90° apart on the housing, while holding the starter assembly in place. The pin should be free to rotate. If it

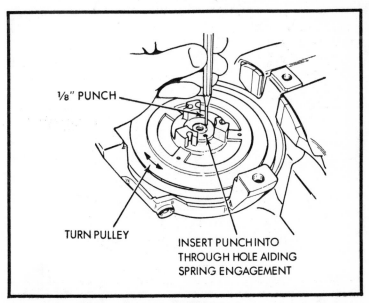

Fig. 6-17 Installing pulley over spring end—larger engines (Courtesy Tecumseh Products Co.)

binds, repeat the procedure. You may have to replace the baffle plate if it is seriously warped.

Eaton starters for large engines are somewhat more complex than the ones we have already discussed. Special assembly procedures may be required, depending upon the application.

Figure 6-17 demonstrates the easy way to mate the main-spring end with the pulley hanger. Insert a small punch in the hole just off pulley center, and rotate the pulley. The punch will guide the spring into the pulley. The next drawing (Fig. 6-18) illustrates the method used to obtain preload. Vise-Grips are held in reserve to clamp the pulley while threading the rope. Preload to approximately 7 turns.

Place a thin, ¾ in. OD (outside diameter) washer over the end of the rope between it and the housing. Tie a knot and melt the frayed end in an open flame. Holding the pulley with thumb pressure, release the pliers. The rope will recoil on the pulley sheave. (Refer to Fig. 6-19.)

After some experimentation with spur-gear drive, Tecumseh has settled upon the vertical-pull starter shown in Fig. 6-20. Servicing for the early type is not radically different than for any other rewind starter. Release the spring preload

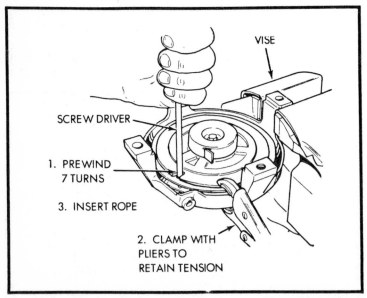

Fig. 6-18 Preloading the larger starter (Courtesy Tecumseh Products Co.)

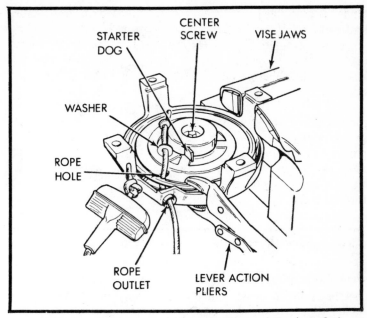

CENTER
SCREW

STARTER
DOG

VISE JAWS

WASHER

ROPE
HOLE

ROPE
OUTLET

LEVER ACTION
PLIERS

Fig. 6-19 Installing the rope (Courtesy Tecumseh Products Co.)

by placing the rope in the slot on the edge of the pulley, allowing the spring to unwind slowly. Inspect the hook-shaped brake spring for wear and distortion. Lightly grease the gear train.

The starter now in use represents a departure from convention in some respects. The pulley moves laterally as it

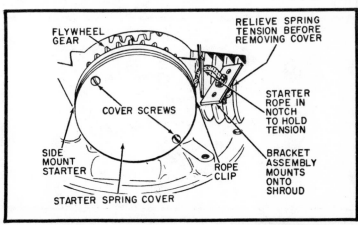

FLYWHEEL
GEAR

RELIEVE SPRING
TENSION BEFORE
REMOVING COVER

STARTER
ROPE IN
NOTCH
TO HOLD
TENSION

COVER SCREWS

SIDE
MOUNT
STARTER

ROPE
CLIP

BRACKET
ASSEMBLY
MOUNTS
ONTO
SHROUD

STARTER SPRING COVER

Fig. 6-20 Vertical-pull starter (Courtesy Tecumseh Products Co.)

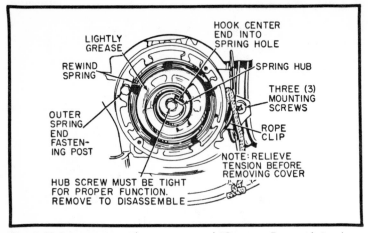

Fig. 6-21 Same starter with cover removed (Courtesy Tecumseh Products Co.)

rotates to engage teeth on the bottom rim of the flywheel. The notched plate in the left side of Fig. 6-20 is a mechanic's aid. It holds rope tension while the mechanic positions his hands against the cover to brake preload. After the rope is swallowed and the spring is fully relaxed, remove the cover screws. Check the hub screw; if the screw has loosened, the pulley will spin idly.

In most cases this is as far as you will have to go. However, rope and engagement problems require separation of the gear and pulley. Remove the ring and the steel thrust washer (Fig. 6-22). The brake spring should be replaced as a

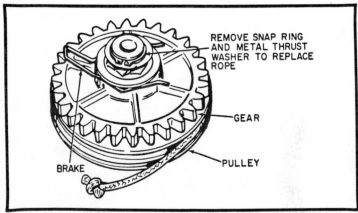

Fig. 6-22 Pulley and gear assembly (Courtesy Tecumseh Products Co.)

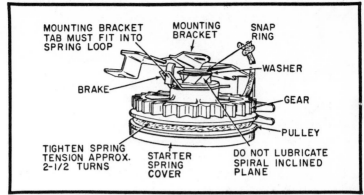

Fig. 6-23 Mounting bracket and brake (Courtesy Tecumseh Products Co.)

matter of course. Clean the brake spring groove and the spiral threads. This starter can be damaged by enthusiastic lubrication. Lubricate the edges of the main spring and the pulley shaft—*not the spiral threads nor the brake spring groove.* Lawnmowers operate in clouds of dust; oil or grease on the exposed parts will hasten wear and may cause erratic action.

Note the position of the brake spring, washer, and ring (Fig. 6-23). When assembling, torque the hub screw securely, and preload the main spring approximately 2½ turns. The mounting plate has elongated holes for gear lash adjustment. There must be $1/16$ in. clearance between the base of the flywheel teeth and the outside perimeter of the starter gear (Fig. 6-24). Failure to set the lash properly can result in a starter explosion—when the engine comes to life.

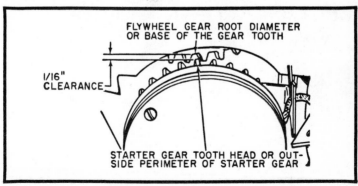

Fig. 6-24 Gear lash adjustment (Courtesy Tecumseh Products Co.)

OUTSIDE RATCHET

Fig. 6-25 Pawl engagement for Lawnboy engines. Unlike most other designs, the pawl engages the outside circumference of the flywheel cup (Courtesy OMC)

And finally, do not attempt to crank the engine without the blade or some other load attached; because of the gear reduction, these starters are hell on kickback.

OMC

OMC has used four different starters on their Lawnboy 2-cycles. The side-pull designs are quite conventional, except that the pawl engages stops on the external diameter of the flywheel hub. When assembled, the pawl should look like that in Fig. 6-26. Preload the main spring moderately, just enough to positively retract the rope. Two turns is the maximum allowed. Inserting the rope through the handle is easier if you use a length of copper tubing as a guide. The handle anchor is

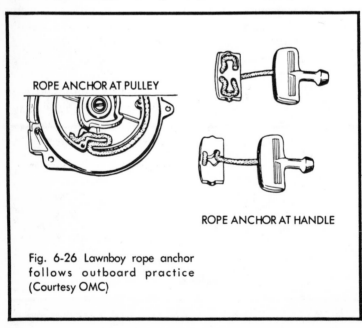

ROPE ANCHOR AT PULLEY

ROPE ANCHOR AT HANDLE

Fig. 6-26 Lawnboy rope anchor follows outboard practice (Courtesy OMC)

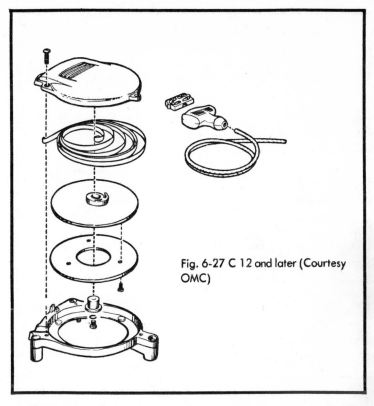

Fig. 6-27 C 12 and later (Courtesy OMC)

tricky. OMC follows outboard-motor practice and secures the rope by a double loop through the inner handle (Fig. 6-26). Be sure to route the rope as shown.

Figure 6-27 shows the series C-12 (and later) in exploded view. (C-12 springs mount as shown in Fig. 6-28.) The C-10 is the design in Fig. 6-28. This starter was distinguished by two consecutive production changes: The first anchors the rope to

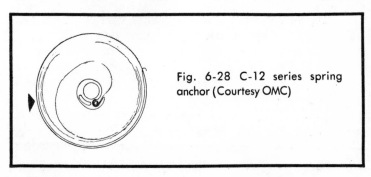

Fig. 6-28 C-12 series spring anchor (Courtesy OMC)

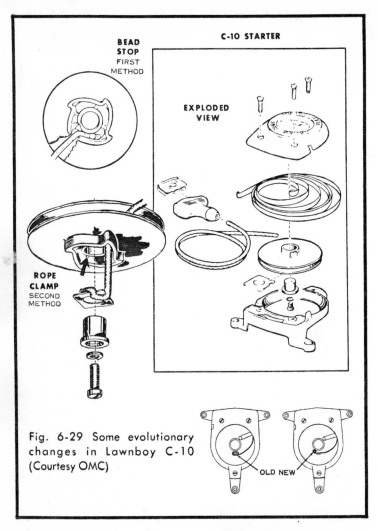

Fig. 6-29 Some evolutionary changes in Lawnboy C-10 (Courtesy OMC)

the pulley by means of a bead stop; later versions had a rope clamp, which proved superior. The clamp can be used on all starters so long as the pulley center is ground down $3/_{64}$ in. to prevent binding. The second change involved the main spring. New springs are secured beyond the mounting stud to give protection against "snapback" breakage.

Main-spring installation is reminiscent of the Briggs & Stratton approach. The spring is pulled into the housing by the pulley. Secure the looped end to the top side of the pulley. Run the free end out of the slot on the starter housing. Wind the

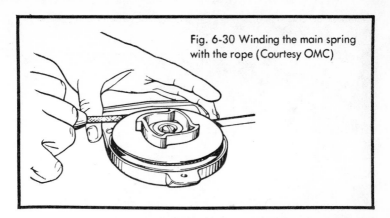

Fig. 6-30 Winding the main spring with the rope (Courtesy OMC)

rope over the pulley so it can be rotated, and install on the housing with the bearing, cap, and screw. Slowly pull the rope—the spring will slide through the slot on the housing until the hooked end mates with a ledge cast into the housing wall (Fig. 6-30).

The 5250 is a skeletal version of the C-12—first time you see one, you think some of the parts are missing. There is no starter cup as such—the spring is anchored to a bracket. When servicing the starter, disconnect the handle and tie the free end of the rope to the bracket to prevent full retraction and loss of preload. Then remove the shroud from the engine; the starter bracket screws are released from the underside.

To install the spring, wind as shown in Fig. 6-31; then turn twice again against spring tension. Knot the rope around the bracket and assemble the shroud to the engine. The rope passes through a wire eyelet on its way out of the shroud. Inspect the guide bushing for burrs which could abrade the nylon rope.

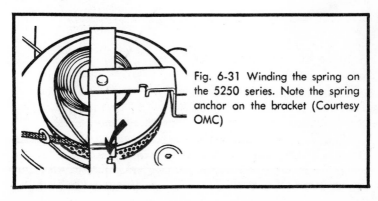

Fig. 6-31 Winding the spring on the 5250 series. Note the spring anchor on the bracket (Courtesy OMC)

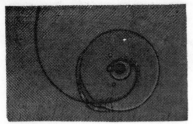

Fig. 6-32 Spring anchor for the vertical-pull D series (Courtesy OMC)

The D series starter is a vertical-pull design, not overwhelmingly different than the Briggs or Tecumseh. Disarm by removing the handle and allowing the rope to retract in a controlled fashion. One screw holds the starter to the engine. The grooved side of the pinion gear goes next to the pulley. Anchor the spring as shown (Fig. 6-32), and guide the free end through the slot in the housing. It is not practical to wind the spring with the starter on the engine. Instead, mount the pulley assembly in a vise, protecting the shaft with wood blocks. Wind two or three times with the rope until the spring anchors. The starter pinion gear operates by means of a brake spring. When replacing this very vulnerable item, spread the ears just enough so the spring snaps into the grooves on the pinion gear. Figure 6-33 illustrates the correct assembly, and

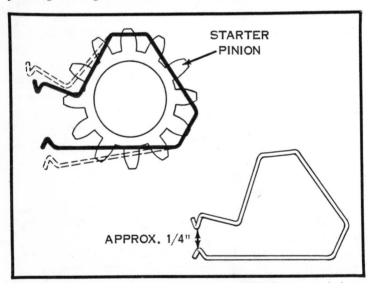

Fig. 6-33 Slide the spring over the pinion, opening the ears as little as possible. Much more than 1/4 in. between ears clearance when disassembled means the spring should be replaced (Courtesy OMC)

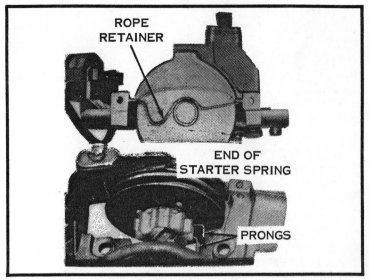

ROPE RETAINER

END OF STARTER SPRING

PRONGS

Fig. 6-34 Assembly reference picture — D series (Courtesy OMC)

the dimension that determines the demise of used springs. If the end clearance is much more than ¼ in., replace the spring. It has warped or fatigued.

Although difficult to show graphically, the ears of the brake spring sandwich the armature-plate ledge when assembled to the engine. That is, one ear is on top of the plate (next to the flywheel), and the other is below it. The main-spring is mounted so that its outside end mates to a slot in the top of the armature plate. The end of the spring is at 12 o'clock. The rope retainer wire has its dropped section facing the coil. See Fig. 6-34 for these relationships. Do not be discouraged if you have to disassemble the starter several times to get it right—mechanics have their problems with this series, too.

IMPULSE STARTERS

Impulse starters are known by a variety of names; they are called ratchet starters, wind-up starters (very appropriate), and speedy starters (wildly optimistic). Whatever the name, these starters work by winding a heavy spring with a crank. The spring is on a ratchet, and remains wound until released.

Many of these starters have been fitted to Briggs and Lauson engines. Clinton has, for the most part, shied away from them. The Briggs is probably the best of the lot, although it is by no means a masterpiece of engineering. Some of the early Tecumseh starters (ca 1959) drove mechanics to drink trying to keep main springs in them. The situation has improved since then, however.

The rationale behind the impulse starter is the reduction of cranking effort; or more exactly, to spread the effort out over time. They are intended for those who may not have the strength to spin a lawnmower engine rapidly enough to start it. The impulse starter can be wound at leisure and (in most examples) released remotely, so that hands and feet are clear of the machine.

The concept looks good on paper (and on the showroom floor), but leaves something to be desired from the mechanic's point of view. In the first place, these starters are not particularly easy to wind. A recalcitrant engine will bring you to your knees. Of course, engines are supposed to start on the first or second impulse, but even the best mechanics find themselves doing the puffing-and-cranking bit. Part of this is because the starter is so inflexible. It's "all systems go" or nothing.

A more serious complaint is deterioration of the spring as the mower ages. Deterioration is more serious in those mowers which have had starting problems, either because of design (e.g., some of the "big wheel" rotaries) or partial system failures. Since the spring is already wound to the point of coil binding, the only alternative is to replace it with another spring. In rare instances you may find that engine compression is too high because of carbon buildup. And while main springs seem to be made of better steel than was used formerly, sudden breakage is by no means unknown.

Don't be surprised to find problems in the clutch mechanism; in most cases, the engagement clutch is a beefed-up version of the clutch used on rewind starters. Impact does not treat it kindly. The ratchet is another trouble spot; tremendous forces are concentrated on the tip of the ratchet—far in excess of even those in a truck transmission. Wear can be a serious problem.

Be extremely cautious when servicing these starters—*The spring can break your arm*. Although it is impossible to foresee all hazards, the following list will give you an idea what to watch for:

204

1) If the spring retainer housing is broken, *expect the starter to explode.* Do not touch it with anything shorter than a 10-foot pole.

2) Do not remove a cocked starter from an engine unless you are forced to do so. When the engine seizes, the natural impulse is to wind the spring and try again. The result is a very dangerous situation. The starter cannot unwind until the engine is free. If it is removed from the blower housing, it will pivot violently on the last bolt until the spring runs down. Once the trigger is released, there is no going back.

How do you cope with these problems? Mechanics are paid a miserly pittance to perform dangerous jobs like this, but have workman's compensation insurance. An amateur takes his chances and pays his own hospital bills. Assuming that you are dealing with a rotary lawnmower, disconnect the plug wire and check the oil. If the sump is dry, add kerosene to (hopefully) free the rod. Prop the machine against a wall, cylinder head up. Hit the blade with a length of two-by-four in the direction of rotation. The impact will, in most cases, free the engine and allow the starter main spring to uncoil. If it doesn't, try a sledgehammer.

If no means of persuasion is effective, the starter will have to be removed—live! *There is no safe way to do this;* all we can do is try to *minimize* the possibility of injury, and take heed of the danger involved. Use several feet of extension bar attached to a speed wrench; climb up on a bench with the modified tool, and begin to remove all the bolts from the mower (on its wheels), *except the last one*—LOOSEN IT ONLY A FRACTION OF A DEGREE. The starter will behave like a helicopter for a few seconds, but shouldn't hurt you if you remain perched on the bench. Other mechanics may have found a better solution to this problem. I would like to hear from them if they have.

The procedure is much safer with Briggs & Stratton starters. Disarm by holding the handle firmly in one hand, to prevent it from turning, and removing the Phillips-head screw (Fig. 6-35).

The spring is dangerous even *after* it is released. Unless the manufacturer allows it, *Do not remove the main spring from its retainer*—discarded springs, coiled in their retainers, are dangerous. The best way to neutralize them is to heat the spring coils around their circumference with a torch. Some

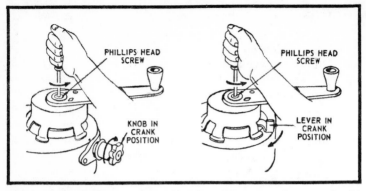

Fig. 6-35 Disarming Briggs impulse starter. Other manufacturers should copy this feature (Courtesy Briggs & Stratton)

residual tension remains, but at least you have not handed the junkman a bomb.

Whenever working around these starters, wear safety glasses and welder's gauntlets.

If, after realizing all the risks, you still want to tackle an impulse starter, the following information will be useful.

Briggs & Stratton

The Briggs starter is one of the simplest and safest of the bunch. The spring and retainer are secured by tabs which may be raised with a screwdriver or a homemade tool (Fig. 6-36). Do *not* attempt to remove the main spring from its retainer. Inspect the ratchet and gear mechanism for wear and replace as indicated. The earlier design used the flywheel fins as holding points; flywheel damage could occur if the control were in the **crank** position while the engine ran. The current design employs a conventional ratchet and pawl. The ratchet wheel is integral with the main-spring retainer and must be replaced as a unit.

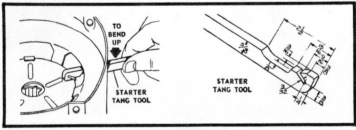

Fig. 6-36 Starter tang tool (Courtesy Briggs & Stratton)

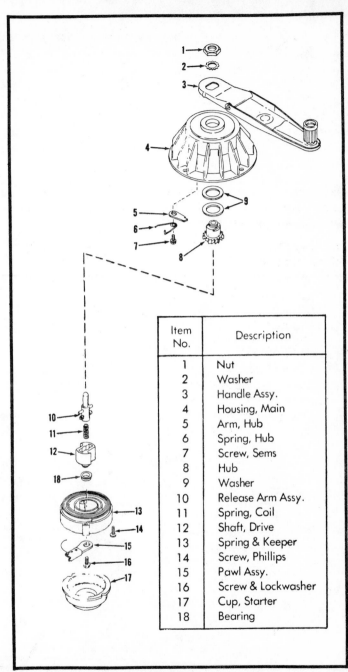

Item No.	Description
1	Nut
2	Washer
3	Handle Assy.
4	Housing, Main
5	Arm, Hub
6	Spring, Hub
7	Screw, Sems
8	Hub
9	Washer
10	Release Arm Assy.
11	Spring, Coil
12	Shaft, Drive
13	Spring & Keeper
14	Screw, Phillips
15	Pawl Assy.
16	Screw & Lockwasher
17	Cup, Starter
18	Bearing

Fig. 6-37 Acme starter (Courtesy Tecumseh Products Co.)

Tecumseh

Tecumseh Products has used several inpulse starters, beginning with the Acme. This unit can be identified by the star-burst pattern formed by the aluminum housing supports (Fig. 6-37). Some are still encountered, although their days are numbered. Check the spring keeper (part 13) for cracks prior to disassembly. *Under no circumstance attempt to remove the spring from its keeper.* In addition to the main spring, high-wear items include the ratchet (5) and the pawl assembly (15). Acme parts are scarce today, but not impossible to come by. Check with your local Lauson distributor, and with out-of-the-way lawnmower shops. An alternative is to replace the unit with a more modern design, although this may entail replacing the hub as well.

The "self-starter" may have either a round or rectangular housing assembly in cast aluminum (part 20, Fig. 6-38). It can

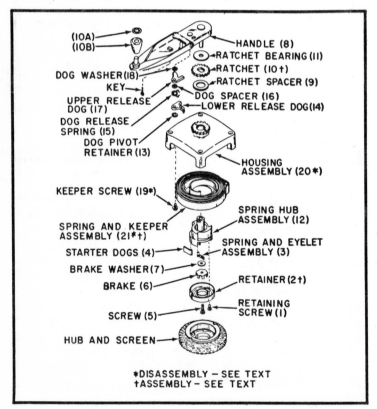

Fig. 6-38 Self-Starter (Courtesy Tecumseh Products Co,)

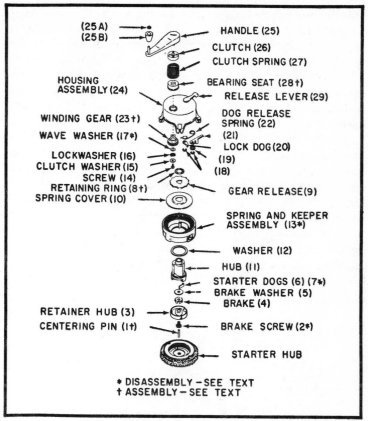

Fig. 6-39 Ratchet starter (Courtesy Tecumseh Products Co.)

be further identified by the fixed gear on top of the housing. The main-spring and keeper are a single assembly, not to be parted for any reason.

To disassemble, remove the four screws which secure the starter to the blower housing. Then, remove the four keeper screws—but only if the spring must be replaced. The keeper and starter housing are an interference fit. Break the parts loose by knocking the housing feet against a hard surface. Be sure the keeper and spring hub assembly (21) have fallen out before lifting the housing. This procedure will give you a modicum of protection from a flying spring.

Install the spring hub assembly before placing the keeper and spring in the housing. Assemble the ratchet so that its teeth are opposed to those in the fixed gear. Be sure the spring (3) hooks to its mounting post. Lubricate the spring assembly with high-pressure grease, and oil the more lightly loaded parts.

The ratchet starter can be identified by the side-mounted release lever and the offset crank (Fig. 6-39). It features a reduction gear between the spring and handle, making it one of the easiest to crank. The spring and keeper (13) are one unit, not to be separated unless one has an overpowering death wish.

The brake screw (2) has a left-hand thread. The clutch assembly is similar to the Eaton pattern; some parts are even interchangeable. The wave washer (17) flattens in service and should be replaced during teardown. The retainer ring (8) should be pressed flat. Use the retainer hub (3) as a driver. The bearing seat (28) should not be greased. After the starter is assembled, drop the centering pin (1) into the crankshaft center hole.

The *Sure Lock* starter (Fig. 6-40) features a pressed steel housing. The handle mounts centrally and the housing is

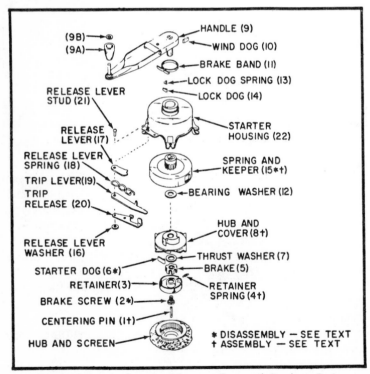

Fig. 6-40 Sure Lock starter (Courtesy Tecumseh Products Co.)

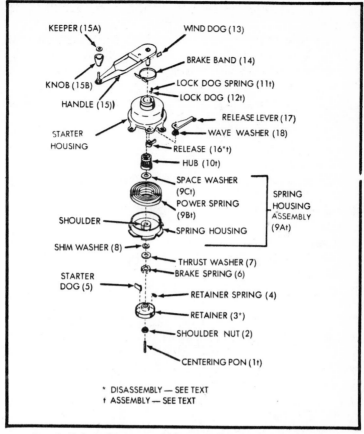

KEEPER (15A)

WIND DOG (13)

BRAKE BAND (14)

KNOB (15B)

LOCK DOG SPRING (11t)

HANDLE (15))

LOCK DOG (12t)

RELEASE LEVER (17)

STARTER HOUSING

WAVE WASHER (18)

RELEASE (16*t)

HUB (10t)

SPACE WASHER (9Ct)

POWER SPRING (9Bt)

SHOULDER

SPRING HOUSING ASSEMBLY (9At)

SPRING HOUSING

SHIM WASHER (8)

THRUST WASHER (7)

BRAKE SPRING (6)

STARTER DOG (5)

RETAINER SPRING (4)

RETAINER (3*)

SHOULDER NUT (2)

CENTERING PON (1t)

* DISASSEMBLY — SEE TEXT
† ASSEMBLY — SEE TEXT

Fig. 6-41 Sure Start starter (Courtesy Tecumseh Products Co.)

splayed with slotted mounting bosses. These recognition points are important, since the external appearance of this starter is similar to the configuration shown in Fig. 6-41. Most internal parts are not interchangeable. The *Sure Lock* main spring is sealed in the retainer.

To disassemble the Sure Lock, remove it from the blower housing. The clutch assembly is similar to all Tecumseh starters, and has a left-hand brake screw (part 2). Lubricate the spring and keeper (15), and the hub and cover (8) with light grease. Use the centering pin (1) as an assembly aid. Pull it about a third of the way out of the screw to engage the crankshaft center.

Most *Sure Lock* and *Sure Start* service difficulties involve the release lever (17). The lever rounds in service. Replace by

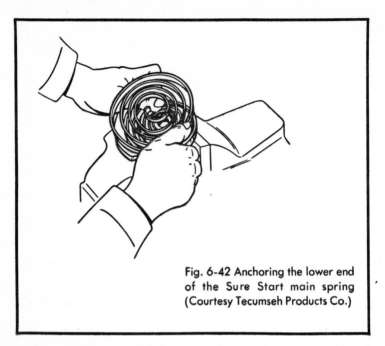

Fig. 6-42 Anchoring the lower end of the Sure Start main spring (Courtesy Tecumseh Products Co.)

cutting out the rivet (21) with a grinding wheel. (Be careful not to elongate the hole.) Install a new lever with a high-tensile bolt, washer, and locknut.

The Sure Start is the newest of the line, and is more reliable than most of the others. If a starter needs replacement, it might be wise to specify one of these, if only to make parts replacement easier in years to come.

Disassembly is not markedly different than that detailed for previous models, except that the spring can be removed from the housing (if necessary) and rewound on the bench. Place the spring housing in a vise (Fig. 6-42). Anchor the spring and engage the free end in the crank as shown Fig. 6-43. Turn the hub with the handle while exerting downward pressure on the handle (Fig. 6-44). Ease the coils over the housing with your (leather-gloved) fingers. With the spring in place, allow the handle to slowly unwind and retract the hub. Place the spacer washer (9C, Fig. 6-41) over the coils before starter assembly.

Clinton supplies several impulse starters. One type is identified by its pressed-steel housing and is not repairable in any real sense of the word. Die-cast starters come in several variations including one with a low profile. These starters are

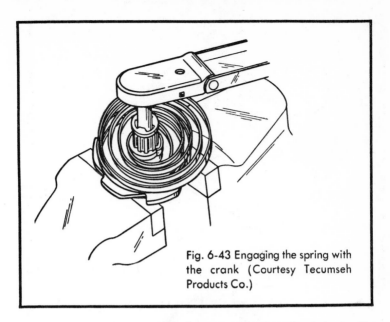

Fig. 6-43 Engaging the spring with the crank (Courtesy Tecumseh Products Co.)

gear-driven and have integral main springs and retainers. To disassemble, remove the four Phillips screws (part 10 in Fig. 6-45), holding the bottom cover to the starter housing. Put the housing on its feet and, holding it at arm's length with your

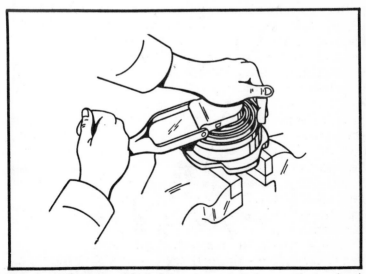

Fig. 6-44 Cranking while holding the spring down with one hand (Courtesy Tecumseh Products Co.)

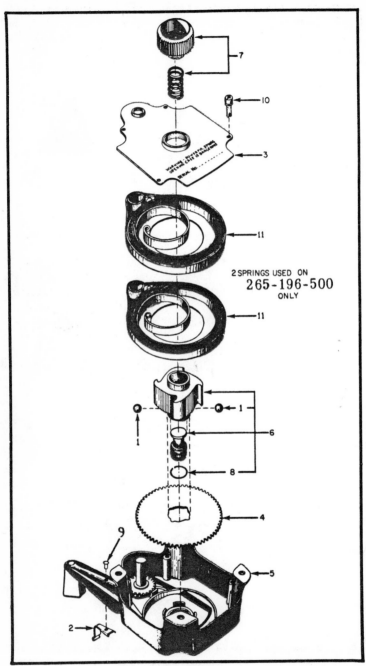

Fig. 6-45 Clinton die-cast starter (Courtesy Clinton Engines Corp.)

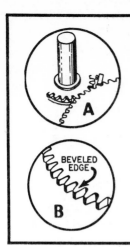

Fig. 6-46 (A) Pawl engagement; (B) beveled edge of the larger gear is toward the housing opening (Courtesy Clinton Engine Corp.)

fingers on the housing and not under it, rap the housing on a clean bench. The spring, retainer, and bottom cover will drop free. Be careful not to dislodge the spring from its retainer. Clean the parts in solvent, and grease with high-temperature Lubri-Plate (500°F melting point). Assemble the pawl and gear as shown in Fig. 6-46.

Chapter 7
Electric Starters and Charging Systems

Lawnmower engines may generate as much as 15 amperes of current in addition to their usual tasks of turning the blade and providing propulsion. This current is needed for a variety of purposes. Some of the larger engines use a portion of it to operate the ignition system; a number of riding mowers have lights and other current-consuming acessories. The primary purpose is to charge the battery. The battery, in turn, provides current to the starter motor.

We will discuss two charging systems under separate headings because the hardware is so different: direct-current systems, and rectified alternating-current systems.

ALTERNATING-CURRENT SYSTEMS

All of these systems have, at their heart, a charging coil mounted under the flywheel on the stator plate. This assembly is known as the *alternator*. The ignition system may be a conventional magneto, battery, and coil; or it might be one of the more modern capacitive-discharge types. In either case, it is entirely independent of the charging system. One will function while the other is inoperative. A starter motor is optional, although it is unusual to find a charging system on a lawnmower without a starter. The convenience far outweighs the relatively small additional cost. Figure 7-1 illustrates the basic circuit. Current leaves the stator plate coils, goes to the rectifier, and from there, to the battery. In this particular example, the rectifier includes a regulator circuit. Small 3- or 5-amp alternators are not regulated.

ALTERNATORS

The alternator consists of a series of coils wound on the stator. Voltage is produced in these coils as magnets on the

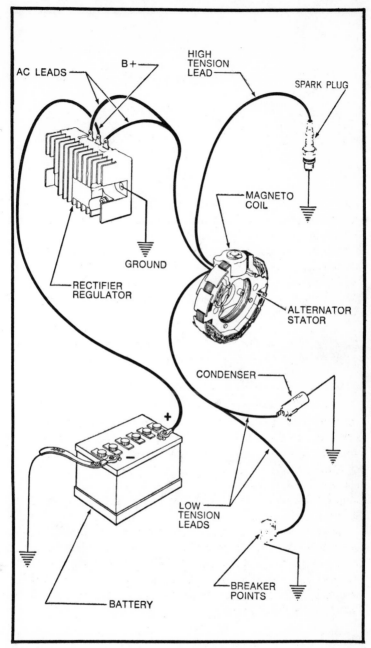

Fig. 7-1 10A ac system with magneto ignition (Courtesy Kohler of Kohler)

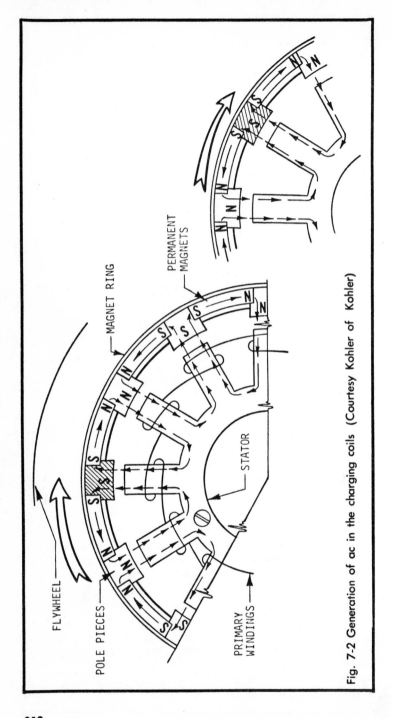

Fig. 7-2 Generation of ac in the charging coils (Courtesy Kohler of Kohler)

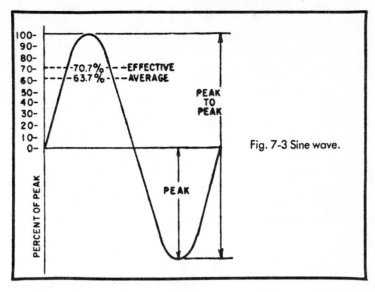

Fig. 7-3 Sine wave.

flywheel (or on a rotor) pass over them. Whenever magnetic lines of force cut through a conductor, voltage and current are induced in that conductor. Look at the stator winding (at 12 o'clock) in the left side of the drawing. The magnetic south pole is directly over it. Magnetic lines of force, which are said to flow from north to south, move upward through the coil. On the right the same winding is shown a few degrees of flywheel rotation later. It is subject to flux reversal. That is, the direction of the magnetic lines of force has changed as the north pole comes into proximity. The direction of current in the conductor is the function of the direction of flux. As it reverses, so does the current.

Of course, this reversal does not happen instanteously. Coil output is in the form of a sine wave (Fig. 7-3). The alternation above the line represents output in one direction; below the line, in the other direction. The average of all values is 63.7% of the peak in any alternation. But the effective value is 70.7% of the peak. This is the heat value of the output compared to dc. Because of the variations in current, ac is only about 71% as effective as dc (which would be drawn as a straight line) in terms of raw heating ability. The frequency of alternations depends upon the number of magnetic poles and on the number of revolutions they make in one minute. Small-engine alternators typically produce frequencies of 150—200 Hz (cycles per second) at governed speed. The frequency shift is not critical, but the increase in voltage and amperage with increased engine speed can be. Figure 7-4 shows the outputs of a typical lighting coil from 1000 to 7000 rpm. The dotted lines represent a low-resistance load.

Rectifiers and Regulators

Alternating current is fine for lights and other accessories, but it is worse than useless for battery charging. It must be converted into direct current of the proper polarity. As engineers say, the ac must be *rectified*. The rectifier or *diode* assembly used in the Lauson 3-amp system is drawn in Fig. 7-5. The rectifiers almost always are semiconductors, usually made of silicon *doped* impurities. Current is blocked from going through the diode in one direction, but finds low resistance in the other.

A single rectifier in series with the coil output blocks half of the alternations. This expedient is found on some Briggs & Stratton models, but it is by no means the most efficient. With

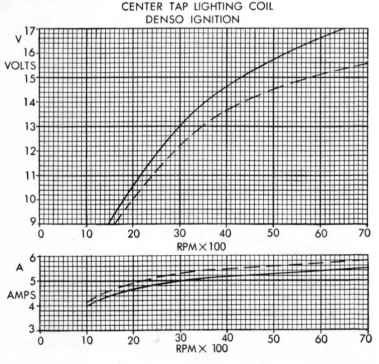

Fig. 7-4 Charging coil

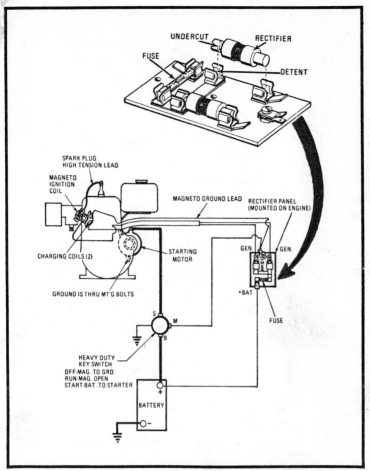

Fig. 7-5 Lauson 3A circuit with rectifier (Courtesy Tecumseh Products Co.)

two or four diodes, and the appropriate circuitry, it is possible to have full-wave rectification. That is, current in both directions is rectified for full benefit of alternator output.

Figures 7-6, -7, and -8 illustrate these circuit alternatives. The barred arrowhead symbolizes a diode. In conventional, negative-to-positive terminology, current actually flows in a direction against that indicated by the arrow. The squiggly line on the extreme left represents the coils. Notice how the output of the half-wave rectifier in Fig. 7-6 consists of pulsating dc. Fortunately, the output frequency is high enough so that lamp filaments appear to burn with a steady glow. The full-wave

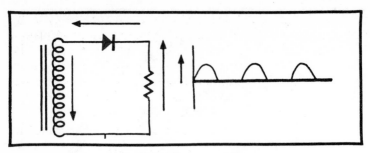

Fig. 7-6 Half-wave rectifier.

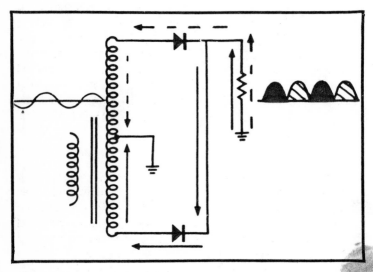

Fig. 7-7 Full-wave rectifier.

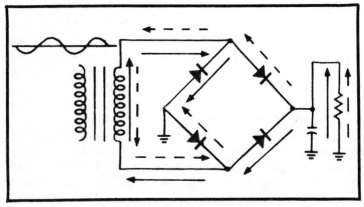

Fig. 7-8 Full-wave bridge rectifier.

rectifier in the next drawing is similar to the one used by Tecumseh. The coil is grounded at the middle (center-tapped). Whatever the direction of current flow in the coil, the ground is always negative. The second full-wave rectifier employs four diodes, arranged in a *bridge* circuit. Bridge rectifiers are found on the larger engines, encapsulated in "black boxes," making it necessary to replace the rectifier as a unit.

Alternating-current systems sometimes employ current and voltage regulation combined with the rectifier circuit. One or more transistors conduct output to ground when battery and charging voltage reach a predetermined value. Low-output systems do not have this sophistication; output is entirely a function of engine speed. The details of these circuits are irrelevant in a book of this type, since no repairs are possible. All that the mechanic can do is to change the unit. A combined rectifier–regulator can be seen in Fig. 7-1. (Note the cooling fins.) Mounting points vary with the manufacturer, but the rectifier–regulator is always easily visible, since it must be well clear of engine heat.

Fuses

Because charging coils and rectifiers are expensive, many engine designers specify a fuse in the output circuit. The fuse is a glass-cartridge automotive type which snaps across two spring clips. The element is a strip of low melting-point metal which warms as current passes through it. A sudden overload will vaporize the element, breaking the connection.

Always replace with a fuse identical to the one originally installed. Obviously a 30-amp fuse will not give much protection in a 7-amp system. The code numbers on the fuse also indicate its reaction time. A really fast fuse blows in 5 microseconds or less. While a fast reaction time is desirable from the point of view of protecting the components, small-engine charging circuits may not be able to tolerate it. These systems are plagued by transients—high voltage spikes—making some time-delay mandatory. A few circuits are so spiky that no fuse can survive them.

A single blown fuse is no cause for alarm. All sorts of factors could be at work; fatigue, high resistance across the element, loose or corroded connections, or transients could cause one to go. But repeated failure can only mean a short circuit somewhere in the system.

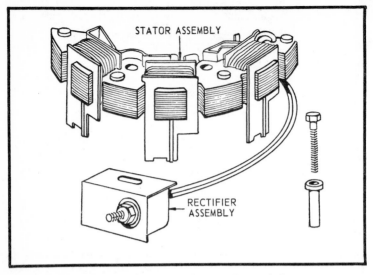

Fig. 7-9 1.5A Alternator (Courtesy Briggs & Stratton)

Briggs & Stratton Systems

The 1.5-amp alternator is designed to feed a small battery. The combination of low charging rate and a small battery is not the best. Unless the complaint is very specific, a wise owner would replace the battery with something more potent—on the order of 24 ampere-hours.

The stator assembly and rectifier box are illustrated in Fig. 7-9. To test for output, run the engine with a No. 4001 sealed-beam headlamp connected between the rectifier output and ground (Fig. 7-10). You can use an ammeter between

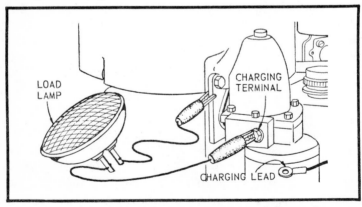

Fig. 7-10 System output test (Courtesy Briggs & Stratton)

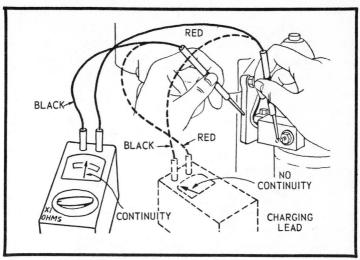

Fig. 7-11 Rectifier check with ohmmeter (Courtesy Briggs & Stratton)

rectifier and the battery's positive post if you prefer. But under no circumstances should output be grounded on this or any other alternator. In other words, the spark test is out.

If the lamp does not light, the fault is either with the stator or the rectifier (assuming the battery terminals are clean and there is no other interruption in the circuit). Test the rectifier first, since it is the more likely failure point. With the engine off, touch the probes of an ohmmeter to the output terminal and ground, as shown in Fig. 7-11. You should get continuity in one direction and high resistance in the other. If not, replace the rectifier box.

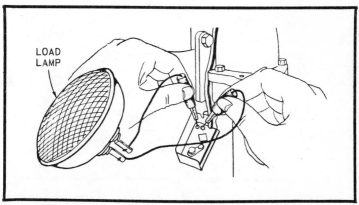

Fig. 7-12 Stator output test (Courtesy Briggs & Stratton)

The stator is tested with a load (lamp) hooked up like the one in Fig. 7-12. Be very careful not to short the leads. Next, check the wires from the stator. They tend to get tangled in the flywheel if they are not routed properly. Before replacing the stator, compare the magnetic strength of the flywheel ring against one known to be good. Failure here is rare, but not impossible.

When installing a new stator, mount the rectifier box on the side of the engine to be sure the leads are properly located. They should be snug against the side of the engine to clear the flywheel. Torque the stator bolts from 18 to 24 in.-lb. Replace the wheel and test.

The 4-amp model is a variation on the one just discussed. It features a full-circle stator with eight charging coils, instead of three, and an inline fuse. The rectifier is mounted under the blower housing. Test points for both the rectifier and the stator are at the connector which can be reached without dismantling the engine.

Troubleshooting procedures begin with a determination of whether the system is shorted. Connect a 12V test lamp between a charged battery and the main wire from the rectifier (Fig. 7-13). Do not start the engine, since this would invalidate the test. After the lamp lights (there is current flow), unplug the connector at the shroud. If the lamp still

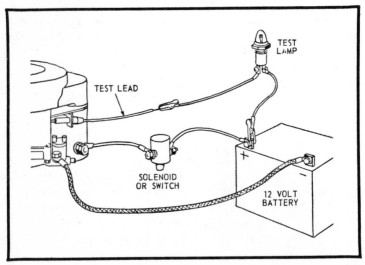

Fig. 7-13 Test lamp connections to detect a short (Courtesy Briggs & Stratton)

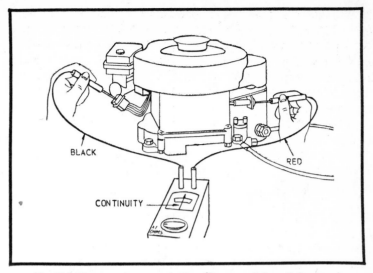

Fig. 7-14 Testing stator continuity (Courtesy Briggs & Stratton)

burns, the rectifier is shorted—if it goes out, the short is in the stator (by elimination).

The half-wave rectifier is fed by a single pin at the connector. Run a resistance test between this pin and a good (clean and unpainted) ground on the engine. Current should flow in one direction but not in the other.

If you suspect that the stator may be open, verify it by touching one probe of your ohmmeter to each of the four pins in the connector, while holding the other probe to the fuse socket (Fig. 7-14). All pins should show continuity. The final test is to connect an ammeter in series with the output.

Briggs & Stratton 7-amp systems may include an isolation diode (Fig. 7-15). The purpose of this diode is to prevent battery discharge through the rectifier assembly when the engine is not running. This system features full-wave, center-tapped rectification. (Refer to Fig. 7-7.)

Check this diode with an ohmmeter. It should conduct in one direction only. Check for grounds with a test lamp as detailed in the discussion on the 4-amp model. If the lamp lights, disconnect the regulator—rectifier plug to isolate these components from the stator, and test the stator for continuity as outlined previously.

The rectifier and regulator boxes, mounted under the shroud, appear identical. The rectifier is closer to the cylinder

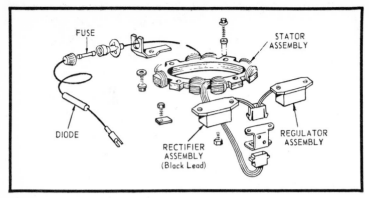

Fig. 7-15   7A alternator with regulator and isolation diode (Courtesy Briggs & Stratton)

head and has two black leads going to it; the regulator has white and red leads. Test the regulator by touching one probe of your ohmmeter to each lead, grounding the other on the shroud. Reverse the leads and repeat the test. The diodes should conduct in just one direction.

The regulator also responds to a simple ohmmeter test. With the meter on the R × 1 scale, connect according to Fig. 7-16. The unit has failed if the red pin shows a reading in either direction; the white pin must show a high-resistance reading in one direction and infinity in the other.

As a final check, connect an ammeter between the output lead and the positive battery post. The meter should show a charge. If the battery is up, load the system with a single No. 4001 headlamp.

The dual-circuit alternator is found on the larger Briggs engines in the 14- to 19-CID range. These engines are intended for riding mowers and garden tractors which have optional lighting circuits. Lighting is a variable load, dependent on the discretion of the user; most designs include a regulator to compensate for the various demands placed on the system. But Briggs engineers were able to eliminate this component by splitting the windings into two independent circuits. The lights receive ac, while rectified dc goes to the battery. The dual-circuit alternator probably costs less to manufacture and certainly makes service work easier.

Figure 7-17 shows the components as they appear on the engine; Fig. 7-18 shows the external circuitry. The lights are not fused, since the windings cannot generate enough current

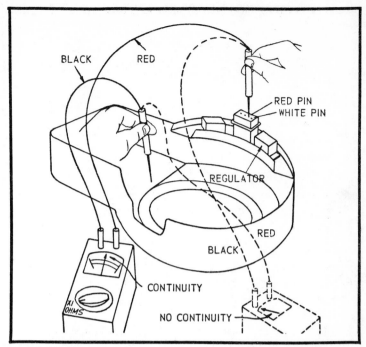

Fig. 7-16 Regulator tests with an ohmmeter (Courtesy Briggs & Stratton)

to burn in the event of an overload. As is the universal practice, the ammeter shunts the starter circuit. Otherwise, the meter would be subject to starter motor drain.

Since two charging circuits are present, each has to be tested separately. The lighting coils have an external ground.

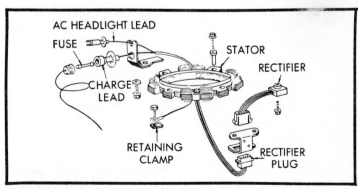

Fig. 7-17 Dual-circuit 5A/3A alternator (Courtesy Briggs & Stratton)

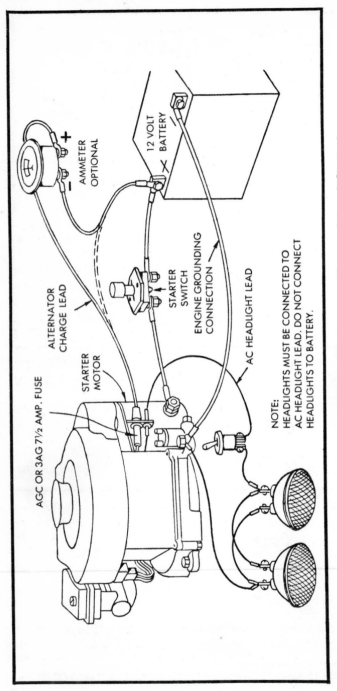

AMMETER
OPTIONAL

12 VOLT
BATTERY

ALTERNATOR
CHARGE LEAD

STARTER SWITCH

ENGINE GROUNDING
CONNECTION

AC HEADLIGHT LEAD

STARTER
MOTOR

AGC OR 3AG 7½ AMP. FUSE

NOTE:
HEADLIGHTS MUST BE CONNECTED TO
AC HEADLIGHT LEAD. DO NOT CONNECT
HEADLIGHTS TO BATTERY.

Fig. 7-18 External circuitry of the 5A/3A alternator (Courtesy Briggs & Stratton)

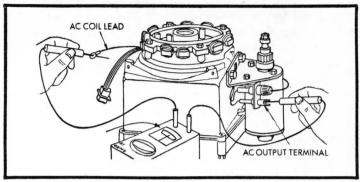

Fig. 7-19 Testing lighting coils (Courtesy Briggs & Stratton)

Check for continuity between the lighting output (located below the fused terminal) and the ground, as illustrated in 7-19. Holding the ground lead away from metal, ground the other probe to the engine block. The ohmmeter should show infinite resistance (unless the windings have shorted). Make the output test with a single No. 4001 headlamp bulb connected to the "hot" lead and to the block. The lamp should flare to full brilliance at medium engine speed; replace the stator if it doesn't. Lighting problems on the machine involve the external circuit. Scrape and retighten all connections and visually inspect the wiring. If necessary, run a point-to-point continuity test with your meter.

The 3-amp battery circuit is tested for shorts with a 12V lamp as illustrated in Fig. 7-13. Disconnect the rectifier plug to isolate the stator, and repeat the test. The charging coils are tested for continuity using the method in Fig. 8-18. Three pins are present; touch the ohmmeter probe to each of the black pins and to the fuse terminal. Repeat the check with one probe grounded. The stator should show continuity and be isolated from the ground. The rectifier has three pins; test each for continuity to ground as described in previous paragraphs. In addition, test between the red pin and the two black pins. The red pin should alternately block and conduct (depending upon the polarity of the meter).

Kohler

Kohler engines are equipped with proprietary systems which follow the normal design pattern for small engines. That is, it shares basic features with those used by Tecumseh, Clinton, Onan, and Wisconsin. The flywheel carries a ceramic ring, fixed by compression and roll pins. Do not attempt to remove this ring; it has been installed at the factory with special tools and was charged before shipment. If the ring has cracked, or if it has lost magnetism with age, the best alternative is to replace the wheel. (A few shops which specialize in magneto work have charging machines, and people with the expertise to operate them.) Protect the ring from shock and remove any metal filings which might adhere during disassembly.

Figures 7-20 and 7-21 are diagrams of two of the 15-amp (15A) charging systems. One employs a conventional flywheel magneto, the other a breakerless ignition. The 30A system, illustrated in Fig. 7-22, is used with a battery-and-coil ignition.

Testing these systems is quite straightforward. The 15A models are tested with the aid of a voltmeter. The B+ terminal should show 13.8V to ground with the engine turning 3600 rpm, without a load. Test regulator action by partially discharging the battery. The lights (if they draw 60W or more) are a sufficient load—or you can bridge the battery terminals with a 100W 2.5Ω resistor. The voltage should increase at the B+ test point. If not, check for a defective stator or rectifier−regulator.

Test the stator by unplugging the ac leads to the rectifier−regulator. If less than 28V is generated at 3600 rpm, replace the stator and windings. If 28V or more is present, the rectifier−regulator is at fault and should be replaced. In most instances, it is impossible to effect repairs on either of these units.

Overcharging will be evident by the ammeter (if so equipped) or by a battery which chronically requires water. Check the B+ output. More than 14.7V means that the rectifier−regulator is not functioning; less than this reading means that the charging system is okay; and by default the measurement points to a defective battery.

The 30A system requires a slightly more complex test procedure. A complete lack of output can involve either the stator or the rectifier diodes. To test the stator, remove the four input leads from the recitifer−regulator. Connect an ohmmeter (R × 1 function) to the two red leads. The meter should register about 2Ω. Make a second resistance reading through the two black leads. Resistance should be in approximately 0.1Ω. You cannot use an ordinary VOM for readings this small. Low-range ohmmeters are commercially

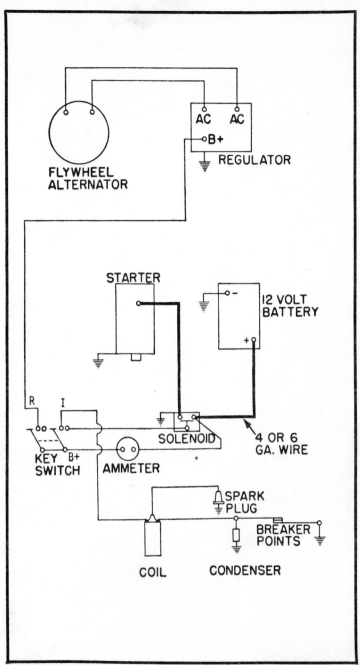

Fig. 7-20 A 15A battery ignition system (Courtesy Kohler of Kohler)

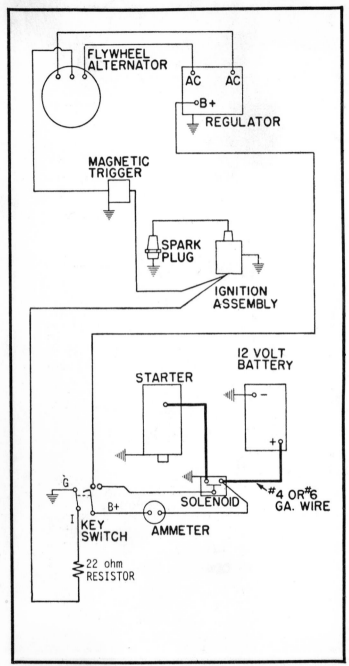

Fig. 7-21 A 15A system—CDI ignition (Courtesy Kohler of Kohler)

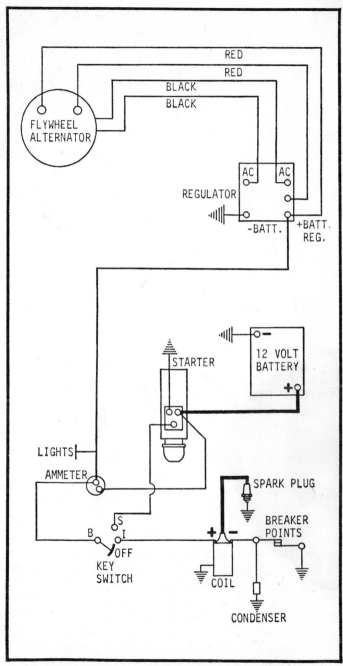

Fig. 7-22 A 30A system (Courtesy Kohler of Kohler)

available, or you can construct your own using precision wire-wound resistors in a Wheatstone bridge. For details consult *Electronic Measurements Simplified* (TAB 702).

To test the diodes, connect an ohmmeter (or continuity tester) between the terminal marked BAT NEG and one of the ac input terminals. Reverse the leads. You should have continuity in one direction and high resistance in the other. Repeat the procedure on the other ac terminal.

A battery which is not receiving a charge (as distinguished from *no output*) is usually a fault associated with the regulator. Bypass it, disconnecting the red lead at the REG terminal. The alternator should charge at rated output. If the ammeter shows significantly less that 30A at highest governed engine speed, replace the stator. Otherwise, replace the rectifier—regulator.

A full charge, with no regulation, may involve the regulator or the regulator winding on the stator plate. Remove the two red leads from the rectifier—regulator and connect them together. Start the engine and run it at full speed. The alternator should deliver at least 4A. Failure to do so means that the stator winding is faulty and must be replaced with the stator. If the stator passes this test, the regulator—by elimination—is at fault.

Tecumseh Charging Systems

The very simple Tecumseh system (Fig. 7-5) is a 3A unregulated design. Output is entirely a function of engine speed. The owner can, however, remove one diode from the rectifier to cut output in half.

The 7A system is identified by the solid-state rectifier—regulator mounted externally, and by the five charging coils on the stator plate. The 5.75A and 10A systems are similar in appearance and function. Troubleshooting procedures differ slightly between the 3A design and those larger. They will be discussed separately.

The smaller unit is first checked with an ammeter connected in series between the rectifier output and the battery per Fig. 7-23. With the engine running at 3600 rpm, the output should be 3A. The second test is made between the output and ground with the battery disconnected. The full-throttle output should be more than 3A. *Note: the engine should not be operated over extended periods with the battery out of the circuit.*

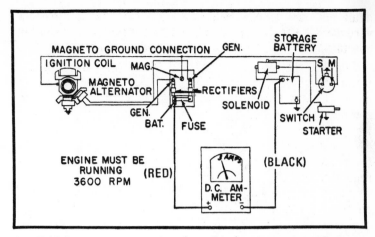

Fig. 7-23 Output test connections for a 3A charging system (Courtesy Tecumseh Products Co.)

Check the diode rectifiers by removing them from the circuit and testing for continuity with an ohmmeter. They should be conductive in one direction and, essentially, open in the other. The fuse, of course, will be conductive in both directions. If the rectifier panel and assorted wiring checks out okay, the stator is the culprit.

Test procedures for the 5.75A, 7A, and 10A units center around the rectifier—regulator. Connect a dc voltmeter (using the 20V scale) as shown in Fig. 7-24; one lead goes to the B+ (positive) terminal wire, and the other goes to ground. Run the

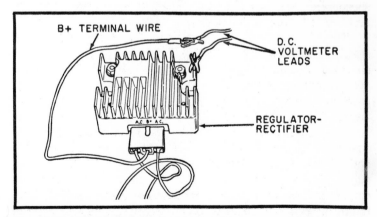

Fig. 7-24 Voltage output test for 5.75A and larger systems—battery out of circuit (Courtesy Tecumseh Products Co.)

237

engine at 3600 rpm. More than 14V output means that the system is functioning normally; more than 0, but less than 14V indicates that the rectifier—regulator should be retired. Zero output may involve either of the two components.

To test the output (to the battery), connect a voltmeter as shown in Fig. 7-25. If over 14.7V, the regulator is defective. Load the battery to drop its terminal voltage below 13.6V. The rate of charge should increase; if it does not, check for a defective stator or regulator.

To determine which is at fault, open the Packard connection at the rectifier—regulator while the engine is running at full governed speed (Fig. 7-26). There should be 20V ac between the output terminals—less is a sign of a defective stator; 20V or more typifies a rectifier—regulator malfunction.

DC SYSTEMS

Prior to the introduction of alternators, small engine charging systems were fed by direct-current generators. A few of these hoary devices survive. Troubleshooting and maintenance procedures are similar to those used with motor—generator combinations, but the parts situation is serious. If the original supplier does not have parts, the best source may be your local auto scrap yard where you might find a matched generator and regulator. Better yet, convert to ac with an alternator and control circuitry from a subcompact car.

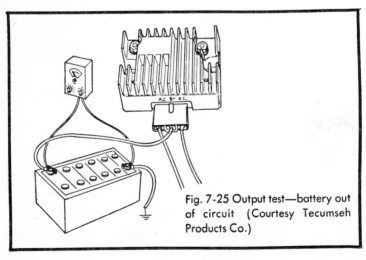

Fig. 7-25 Output test—battery out of circuit (Courtesy Tecumseh Products Co.)

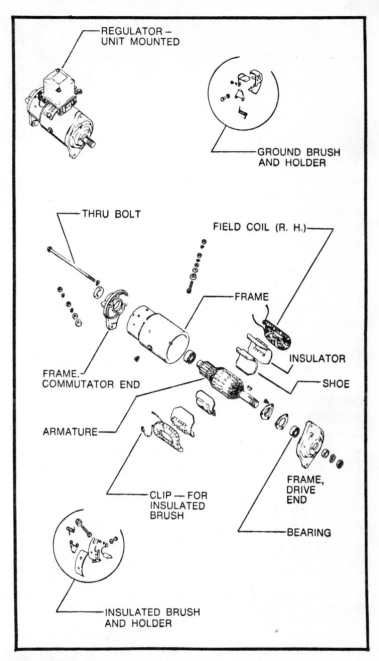

REGULATOR –
UNIT MOUNTED

GROUND BRUSH
AND HOLDER

THRU BOLT

FIELD COIL (R. H.)

FRAME

FRAME,
COMMUTATOR END

INSULATOR

SHOE

ARMATURE

CLIP — FOR
INSULATED
BRUSH

FRAME,
DRIVE
END

BEARING

INSULATED BRUSH
AND HOLDER

Fig. 7-26 Motor — generator in exploded view (Courtesy Kohler of Kohler)

Motor—generators have been used on Kohler and Briggs engines, on several Japanese motorcycles, and on the under-30-hp Chrysler outboards. A dc generator can be converted into a fairly serviceable motor by the application of current to the brushes. Ideally, the motor and generator field windings should be distinct and separate. The motor function draws a lot of current—the reason for the large-diameter turns on the field coils. Motor—generators used for lawnmower and light industrial applications do not have this refinement.

The attraction of the motor—generator is its relative compactness, at least when compared to a separate motor and dc generator. Another advantage is that it remains engaged at all times; there is none of the brutal action associated with Bendix and solenoid clutches.

In the generator function, the armature is rotated by the engine for a 4:1 speed advantage. This overgearing enables the generator to spin fast enough to deliver a charge at low engine speeds (cut-in is at 2000 rpm for most designs) and gives a torque advantage during cranking. Spring-loaded carbon brushes bear against the commutator. The segments (or bars) which make up the commutator are terminals for the individual armature loops. An opposed pair of segments forms a complete circuit. Adjacent segments are insulated from each other and all segments are insulated from the armature shaft. Two field coils are provided, and are wired in series with the armature. The field coils are mounted on iron shoes (shown clearly in Fig. 7-26); when energized, the shoes become powerful electromagnets. The cast-iron frame is part of the magnetic circuit. The motor—generator shown has antifriction bearings on the commutator and drive ends. Other designs employ special bushings at these points.

The field coils generate magnetic lines of force; as the armature rotates, its individual loops cut through these lines of force, developing a voltage which is taken out of the armature circuit by the brushes. One brush is grounded to the frame and the other is insulated. During the motor function, current is fed into the brushes, which energizes both the fields and the armature. Attraction and repulsion between the armature windings and the fields produce torque.

TROUBLESHOOTING THE GENERATOR

If regulator output is zero, first check the ground connection. In many applications the regulator is grounded to the engine block through a woven wire strap; the strap can become resistive over time. Inspect for good brush contact to the commutator, and for signs of arcing. Pitting between adjacent segments usually means at least one open armature loop.

Generator output can be determined as follows:

1. Disconnect the field lead from the regulator (marked F or FD).
2. Connect an ammeter between the B terminal and the B+ post of the battery.
3. Run the engine above 3000 rpm and watch the reading.
4. If the generator delivers more than 2−3A, you can safely assume the fields have grounded themselves to the frame

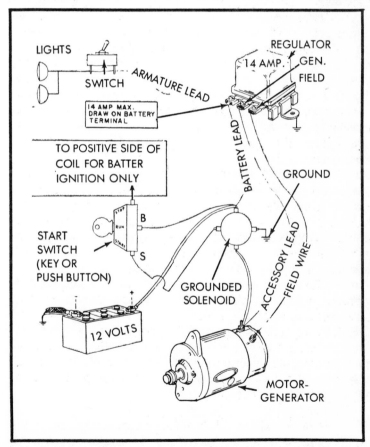

Fig. 7-27 Regulator circuit (Courtesy Tecumseh Products Co.)

5. With all accessories off, momentarily ground the field lead. The generator should deliver its maximum rated charge. Perform this test quickly to prevent overheating.

More elaborate troubleshooting and repair techniques are essentially identical to those used for starter motors and are discussed under that heading.

DC Regulators

Dc systems betray their age by the use of vibrating-reed regulators (Fig. 7-28). These devices are subject to failure from heat and contact oxidation, but do have the incidental advantage of being immune to polarity reversals and high-voltage spikes. Most regulators have two windings—one to control voltage and current, and the other to prevent battery current from spilling back into the circuit after shutdown. (Some readers may be more familiar with automotive regulators which have a separate winding for each function.)

Figure 7-27 shows the wiring connections for a typical regulator circuit. Most regulator problems are the result of poor grounding or dirty contact points. Check all connections for continuity and be particularly scrupulous about the regulator ground. The points can best be restored by burnishing with a riffle file. *Do not use emery cloth or sandpaper.*

To adjust the cutout relay, disconnect the battery and press down on the armature until the points just close. Measure the air gap between the armature and the center of the core. On nearly all current-production units the air gap should be 0.02 in.; vary it by moving the armature on its hinged mounting. Adjust the point opening 0.02 in. by bending the armature stop. The closing voltage is adjusted with the battery in the circuit. It is controlled by the adjustment screw shown in Fig. 7-29. The closing voltage should be at least 0.5V less than the operating voltage.

The current—voltage relay requires two adjustments: armature air gap, and voltage setting. To check the air gap, push the armature down until the contact points are still just touching, then measure the air gap. It should be 0.075 in. (see Fig. 7-30). Be sure the points are aligned and the screws are tight before going to the voltage setting.

The voltage setting is controlled by a hanger or screw. Run the engine at governed speed and adjust for a voltmeter reading of 12.8V taken between the B+ terminal and ground.

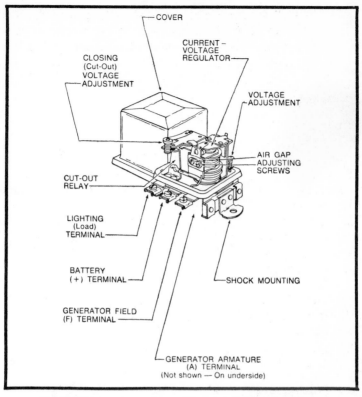

Fig. 7-28 Two-unit regulator (Courtesy Kohler of Kohler)

After each setting, replace the cover and allow the engine to run for a few minutes to stabilize the voltage and temperature before rechecking.

The specifications given are specifically for Kohler engines, but generally apply to most all others. If you are in doubt about the exact data, request it from the regulator manufacturer, citing the type number. Voltage output values are subject to some individual interpretation. Many mechanics adjust the regulator higher than 12.8V, especially for summer operation.

After the generator or regulator has been serviced, it is necessary to *polarize* the generator. This is done by momentarily connecting a jumper between the B terminal of the regulator and the A (armature) terminal of the generator.

Regulator failure is rarely an isolated event. Usually, some other part of the circuit is at fault. Burned contacts or

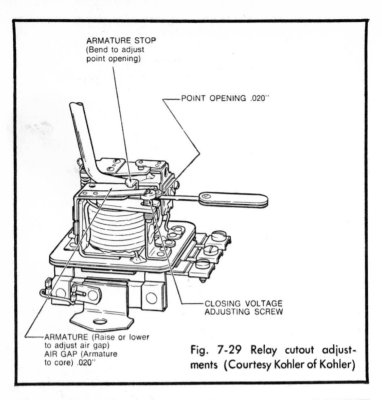

ARMATURE STOP
(Bend to adjust
point opening)

POINT OPENING .020"

CLOSING VOLTAGE
ADJUSTING SCREW

ARMATURE (Raise or lower
to adjust air gap)
AIR GAP (Armature
to core) .020"

Fig. 7-29 Relay cutout adjustments (Courtesy Kohler of Kohler)

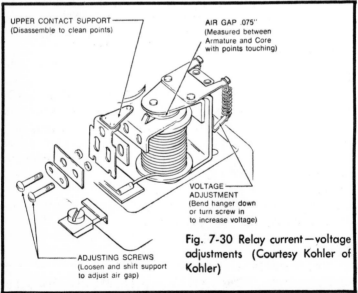

UPPER CONTACT SUPPORT
(Disassemble to clean points)

AIR GAP .075"
(Measured between
Armature and Core
with points touching)

VOLTAGE
ADJUSTMENT
(Bend hanger down
or turn screw in
to increase voltage)

ADJUSTING SCREWS
(Loosen and shift support
to adjust air gap)

Fig. 7-30 Relay current—voltage
adjustments (Courtesy Kohler of
Kohler)

244

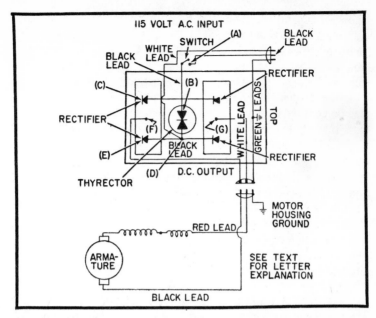

Fig. 7-31 A 110V starter-motor rectifier (Courtesy Tecumseh Products Co.)

springs in the cutout relays are commonly exhibited by units that were not polarized. The same symptoms in the current—voltage section may point to high resistance in the charging circuit or a poor ground. Burnt windings augur extremely high resistance or an open circuit. A short will produce the same symptoms.

STARTING SYSTEMS

In addition to starter—generator combinations, lawnmowers may be equipped with conventional 12V ac or dc generator-fed motors, 12V battery-powered motors, or 110V ac types. The latter operate from household current.

Before assuming the starter is bad, turn the engine by hand to be certain it is free; also, check the battery. Starter engagement periods should not last more than 10 seconds—with 1-minute intervals between bouts. A 15-minute interval should follow every five attempts.

110V Starters

The unique feature of these starters is the rectifier which changes ac line current into 90V pulsating dc. Figure 7-31 shows a typical rectifier in schematic form. The diodes are mounted in pairs on heatsinks. The *Thyrector* is a Tecumseh feature. Normally it is an insulator, but beyond its rated voltage it conducts, thus protecting the diodes from voltage surges in the ac line. The rectifier illustrated can be repaired. Test the diodes for unidirectional conductivity. The thyrector can be verified with a 7.5W lamp in series with a 115V ac power line. If the bulb glows, the thyrector is a dud. Most connections are soldered and you should be very circumspect with the soldering gun. Heat will cause irreversible changes in the diodes and Thyrector. Monitor the temperature holding your finger to the component—as long as you can take it, so can the part. The only test remaining is an output reading across the dc terminals with a voltmeter.

The Briggs & Stratton rectifier is less complicated, but is serviceable only as a complete unit. It should deliver no less than 14V below line current across a 10,000Ω;1-watt resistor (Fig. 7-32).

Figure 7-33 is an exploded view of the four-pole Tecumseh starter. Two-pole American Bosch designs have also been used on Lauson engines in the recent past, and continue to be available in kit form for after-sale installation; the same starter is used by Briggs & Stratton. Free-running speed

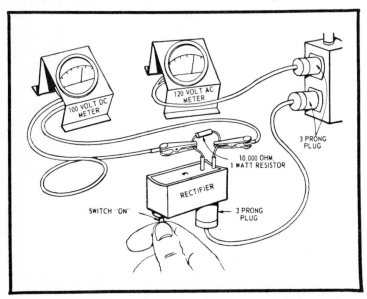

Fig. 7-32 Testing the Briggs rectifier (Courtesy Briggs 8 Stratton)

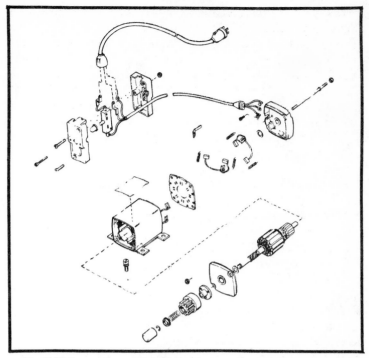

Fig. 7-33 Four-pole starter motor (Courtesy Tecumseh Products Co.)

should be 5200 rpm with a maximum current drain of 3.5A (Fig. 7-34). The four-pole model should draw 3A without a load. During cranking, with all parasitic loads removed, and the engine and oil temperature approximately 70°F, it should be capable of spinning the crankshaft 1000 rpm while drawing no more than 7A. This test should not exceed 10 seconds.

Failure of a starter to meet performance specifications can be traced to:

- Binding (burned or dry) armature shaft bearings
- Worn armature shaft bearings
- Brushes worn below half their new length
- Brushes sticking in their holders
- Shorted, opened, or grounded armature
- Shorted, opened, or grounded field
- Faulty switch or solenoid
- Faulty rectifier (ac models only)

Remove the through-bolts to inspect the brushes and armature. But before doing so, scribe a reference mark on the

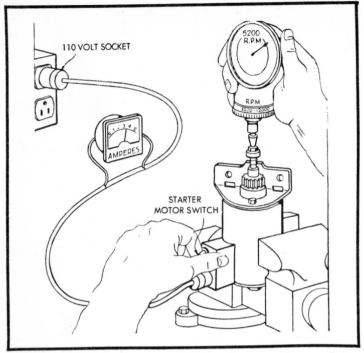

Fig. 7-34 Testing the two-pole American Bosch motor (Courtesy Briggs & Stratton)

end plate. A reddish or brownish discoloration is normal on the commutator bars; it means that the brushes have mated. Glaze and other minor imperfections can be removed with No. 00 sandpaper as shown in Fig. 7-35. An out-of-round condition, or deep scratches or pits, should be corrected on a lathe. After

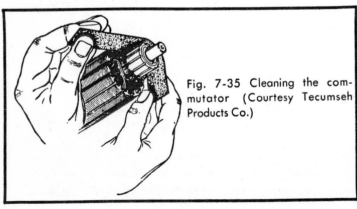

Fig. 7-35 Cleaning the commutator (Courtesy Tecumseh Products Co.)

either of these operations, undercut the mica with a tool designed for the purpose, or with a narrow straight-edged jeweler's file. Be sure the channels are straight and true. Polish the commutator lightly, and clear all metal dust from between the segments with compressed air.

Bearings are the next most likely area of failure. Rotate antifriction bearings by hand. The inner race should turn easily with very little radial play, or none at all. In most starters these bearings are prelubricated and sealed, making it necessary to replace a dry bearing with a factory equivalent. Drive out the old bearing and, using a drill press or a vise, press in a new one (on the numbered side) to the depth of the original. Bushings tend to take on an egg shape in service. Replace them as you would ball bearings, driving out the old ones—being careful not to scratch the inner diameter of the bearing boss—and pressing in the new ones. Bushings on the end covers can be removed by either of several methods; a small chisel can be used to split the bushing. This method is adequate as long as extreme care is exercised not to gouge or otherwise distort the boss.

American Bosch starters have a rear bushing with a face made to accept thrust loads; collapse the face inward, distorting the bushing. Perhaps the best method (and certainly the most elegant) is to employ a punch of exactly motor shaft diameter. Pack heavy grease into the boss and drive the punch down with a hammer. Hydraulic pressure will lift the bushing clear! The job, as far as bushings are concerned, is complete—they do not require finish-reaming for starter service.

The armature can develop openings or shorts in its windings. Check for shorts between the motor shaft and the armature form, with a 120V test lamp. All iron or steel parts must be insulated from all nonferrous (brass or copper) parts. Check adjacent commutator bars by the same method. In general, a shorted armature cannot be repaired since the shorts are almost always internal, and rewinding is more costly than a new part.

Make the usual electrical tests—this time using an ohmmeter between the brushes and their holders on the end plate. Undergrounded brushes should not show continuity to the plate, and grounded brushes should show no resistance.

The next two tests require a "growler." If you don't want to buy one, you can build your own—using the core of a small

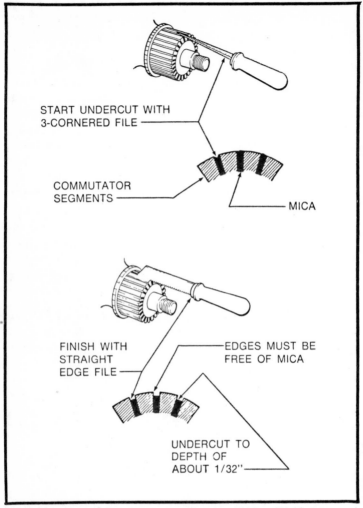

START UNDERCUT WITH
3-CORNERED FILE

COMMUTATOR
SEGMENTS

MICA

FINISH WITH
STRAIGHT
EDGE FILE

EDGES MUST BE
FREE OF MICA

UNDERCUT TO
DEPTH OF
ABOUT 1/32"

Fig. 7-36 Undercutting mica (Courtesy Kohler of Kohler)

transformer. (For construction details see *Small Appliance Repair Guide*, TAB 515. The growler converts the armature into the secondary of a transformer by saturating it with a 60-cycle magnetic field.

If one of the armature windings is internally shorted, a hacksaw blade will vibrate when placed over the affected armature-form segment (Fig. 7-37). And if one or more of the windings is open, you will be able to see sparks between the blade and adjacent commutator segments.

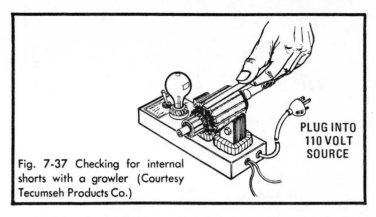

Fig. 7-37 Checking for internal shorts with a growler (Courtesy Tecumseh Products Co.)

Field coils should be studied for signs of burning and for evidence of contact with the armature. The latter is usually indicative of a severely bent motor shaft or loose bearings. Test the fields to ground with a 120V lamp. Resistance checks require a Wheatstone bridge. Since this kind of test gear is considered exotic by most lawnmower repairmen, fields are replaced by default. If everything else checks out okay, it can be assumed that the fields are internally shorted.

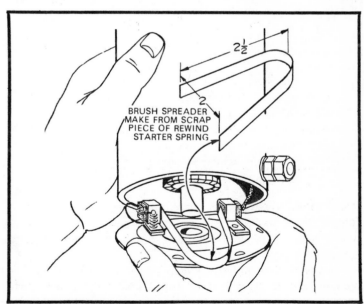

Fig. 7-38 A tool used to hold the brushes apart during assembly (Courtesy Briggs & Stratton)

251

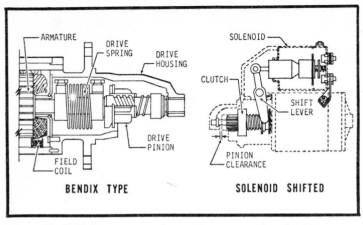

ARMATURE

DRIVE
SPRING

DRIVE
HOUSING

FIELD
COIL

DRIVE
PINION

SOLENOID

CLUTCH

SHIFT
LEVER

PINION
CLEARANCE

BENDIX TYPE

SOLENOID SHIFTED

Fig. 7-39 Starter-motor drives (Courtesy Kohler of Kohler)

Most small starters require a "shoehorn" to slip the brushes over the commutator. Figure 7-38 shows how such a tool is made from a length of rewind spring. Align the end cap with the reference scribe marks you made earlier. Assemble the unit and bench-test it.

Starter Drives

Bendix inertial drives are most popular, although some manufacturers use the solenoid type. These two configurations are shown in Fig. 7-39. The solenoid type functions as a relay to pass heavy currents to the starter, and as a linear motor to move the pinion gear into engagement with the flywheel. The Bendix type has no electrical function; it operates on inertia to move the pinion along a helix and into mesh with the flywheel ring gear. A variation of the Bendix is pictured in Fig. 7-40.

The Bendix should be examined for chipped or excessively worn teeth, and for spring breakage or serious distortion. The gear should move easily on its helix. The helix and associated parts should not routinely cleaned in Varsol or kerosene, but assembled dry. Oil or grease is not required and will only cause untimely failure by sticking, and by attracting airborne abrasives in the form of dust.

The solenoid pinion should be inspected for wear and for proper ratchet action. It is the pinion's role to transmit starter torque to the flywheel; it must be free to overrun when the engine fires. Failure to engage may be attributed to a binding linkage, shorted windings, or open windings. Test

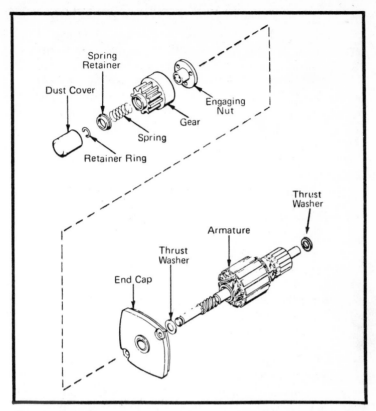

Fig. 7-40 Bendix starter-motor drive in exploded view (Courtesy Tecumseh Products Co.)

specifications are not available from mower manufacturers; solenoids can only be tested by substitution. Obvious faults include failure to engage, machine-gunning, and starter overrun. The latter symptom could involve the brushes.

Relays

Most starting circuits include a relay to switch heavy starting currents. They may be part of the solenoid or a separate unit used in conjunction with a Bendix starter drive. The relay is energized by light current from the pushbutton or keyswitch mounted on the control console. The relay provides a more direct path for electricity from the battery to the motor, which results in less resistance and shorter cables.

Two types of relays are used; some are internally grounded as shown in Fig. 7-5, and the others are externally grounded. The former have three connections—two large terminals for the main switching current and one small terminal to the switch. Externally grounded relays have an additional small terminal which connects to a ground wire.

Relays either fail at their windings (which may open or become shorted), or at the switch contacts (which can melt because of high-current demands). If the motor refuses to crank, jump the large terminals on the relay with a screwdriver blade. Should it now crank, you can be sure the relay is at fault; if it doesn't, some other circuit component is involved.

In general, relays are not repairable—although stubborn mechanics have been known to cut the case apart, file the contacts, and epoxy it back together. Any 12V automotive relay can be used, assuming that you make certain that it is grounded. The best choice is, of course, a factory part.

Batteries

The tie point between the charging and the starting system is the battery. Most lawnmowers use the familiar lead—acid storage battery that was developed almost a century ago, with few improvements since then.

Each cell consists of lead plates sandwiched between electrically neutral separators. Cedar was once used for separator material, but this dense, stable, and warpless timber has become almost commercially extinct. Today, most battery manufacturers employ fiberglass separators. You can still purchase very cheap batteries with slats of fir between the plates, but this material is not recommended. The internal battery components are lead straps between plates which, in modern designs, are under the cover. This location makes it difficult to test individual cells, but it does reduce self-discharge.

The plates are submerged in electrolyte—a solution of distilled water and sulfuric acid. When electrolyte is added, the battery immediately begins to function. What happens involves the higher reaches of chemistry, but can be summarized in a rudimentary manner.

A battery's plates are made of pure lead (Pb) in a spongy form; positive plates consist of lead dioxide (PbO_2) on a grid of lead and antimony. The electrolyte is water (H_2O) and sulfuric acid (H_2SO_4); the usual mix is 32% acid by weight. As the battery discharges, both plates become lead sulphate ($PbSO_4$).

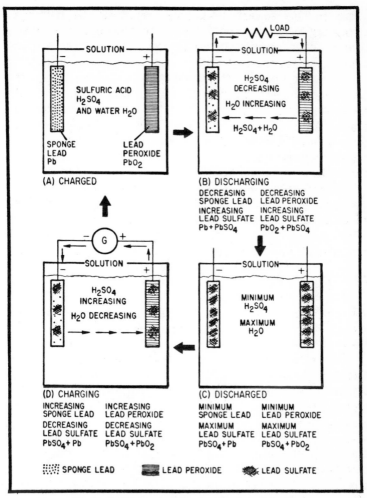

Fig. 7-41 Chemical action in a lead—acid cell.

The percentage of water in the electrolyte increases since the SO₄ radical splits off from the sulfuric acid to combine with the plates.

During charge, the reactions reverse—the electrolyte becomes more acidic, and the plates convert back to lead and lead dioxide. These reactions are shown in Fig. 7-41.

Battery Tests

The most reliable test requires a large rheostat or a carbon pile. This load is adjusted for three times the ampere-hour capacity of the battery, and the battery is discharged for 15 seconds. If at the end of this period the terminal voltage should fall below 9.6V, recharge and test again. Failure to hold a minimum of 9.6V during discharge is grounds for replacement. In a pinch you can use the machine's starter motor—but do not operate the starter for more than 15 seconds. Allow 10−15 minutes between runs for the starter windings to cool.

Individual cells can be tested with a hydrometer which, while by no means the ultimate in test equipment, will give you a good notion of the state of charge. The hydrometer's working principle takes advantage of fact that the electrolyte approaches pure water as the battery discharges. Water has a specific gravity of 1; sulfuric acid has a specific gravity of 1.83. In other words, it is 1.83 times heavier than distilled water. In general, lead−acid batteries are adjusted (by dilution) to a specific gravity between 1.26 and 1.28.

One of the difficulties with hydrometer testing is that it is subject to several variables. The engine should be operated several hours after water is added to the cells to insure complete mixing. Then, it should be parked at least overnight and restarted once or twice to reproduce the average use cycle. Insert the tip of the hydrometer into the cell and draw enough electrolyte into the chamber to set the float adrift. Do not force the hydrometer tip against the tops of the plates when you draw electrolyte. It is possible to bend them and cause internal shorting. Hold the hydrometer level and take a sight across the liquid level to the scale on the float. Do not be misled by liquid which clings to the sides of the chamber and to the float stem. Since the specific gravity will change with the temperature of the electrolyte, it is desirable to plug in a correction factor. Professional hydrometers have an integral thermometer which takes the temperature of the electrolyte as the cell is being tested. However, any accurate thermometer is sufficient. Add 0.004 to the reading for each 10° above 80°F, and subtract a like amount for each 10° below 80°F.

The chart in Fig. 7-42 gives temperature-corrected readings and approximate states of charge. If there is more than 0.05 point difference between the highest and lowest cell readings, replace the battery.

The best way to recharge a battery is to use the machine's charging system. For a rough idea of how long this will take (with accessories off), divide the generator output into the

Battery Condition	Specific Gravity
Discharged	1.110 to 1.130
No useful charge	1.140 to 1.160
25%	1.170 to 1.190
50%	1.200 to 1.220
75%	1.230 to 1.250
100%	1.260 to 1.280

Fig. 7-42 State of charge and specific gravity correlation (Courtesy Tecumseh Products Co.)

ampere-hour capacity. For example, a 24 A-hr battery will require 5 hours at the 5A output, which is characteristic of small alternators. The battery can be removed from the machine and put on a charger; however, the rate should be commensurate with charger output. In no case should the electrolyte temperature be allowed to climb above 125°F. Connect the leads before turning the charger on, and remove them after the charger is switched off. Batteries give off hydrogen gas; the quantity is greatest at near full charge. These precautions along with a ban of smoking and open fires will do much to prevent battery explosions.

Battery Maintenance

Careful maintenance will extend the life of the battery well beyond the 2-year norm for power mowers. Prior to storing the machine for the winter, remove the battery for periodic, once-a-month recharges. A trickle charger will pay for itself several times over the cost of batteries it saves from sulfation. The top of the battery should be kept free of grease, spilled electrolyte, and dust at all times. Batteries can usually be wiped clean with a damp cloth. Stubborn deposits can be removed with a mixture of one part detergent, two parts baking soda, and eight parts water. Do not allow the solution to enter the vents in the filler caps—it will dilute the electrolyte. Rinse with water and dry. Keep the cells filled to the mark (but not overfilled so that electrolyte escapes when the battery is heated). Periodically remove the terminals and scrape to bare the metal. Be gentle with the terminals—do not twist or pry. If one refuses to part, use a battery-terminal lifter. These tools look like small gear pullers and extert force vertically to lift the terminal without stressing the case or strap. After cleaning, retighten and coat the outside of the terminal with grease. The battery mounts should be secure; a loose battery will soon distintegrate its positive plates.

Purchasing a Battery

The rule is to buy the best battery that you can afford and the largest capacity that you can conveniently mount on the machine. Battery capacity is a measure of its energy storage capability and is a function of the care taken in manufacture, the materials used, and the plate area. The quality of construction can only be estimated by the reputation of the builder and by the price. Plate area is rarely mentioned in battery specifications, although you can get some very rough idea from the weight of the battery—the heavier the better. A more scientific approach is represented by the various rating systems.

The most venerable standard is ampere-hour capacity. The battery is discharged continuously for 20 hours until the terminal voltage drops to 10.5V. A battery which can deliver 2A over this period is designated as a 40 A-hr (2×20) battery. This does not mean that the battery can deliver 40A for 1 hour or 160A for 15 minutes, because the output is not linear.

Since batteries are primarily for starter motor operation, other more rigorous ratings have been developed. *Zero-cranking power* is a hybrid combining a voltage reading and a time duration. The battery is chilled to 0°F and discharged at 150A for 5 seconds. The voltage is read across the terminals. Discharge continues until voltage drops to 5V. The time in minutes it takes to reach this point is the second part of the standard. The higher both numbers are, the greater is the capacity of the battery. *Cold-cranking power* is replacing the hybrid standard. Again, the temperature is lowered to 0°, and the load is adjusted to bring the voltage down to 7.2V. Output in amperes is measured for 30 seconds at this reduced voltage. The greater the output, the better the battery. Yet another low-temperature yardstick is *peak wattage* (volts × amps) at 0°F.

These are only some of the rating systems used. None are absolute; nevertheless, taken together they will give you a good picture of relative battery performance.

Nickel—Cadmium Batteries

The nickel—cadmium or *nicad* battery is increasing in popularity with power mower designers. It is the first economically feasible alternative to the lead—acid cell, and has several advantages over its aging relative. In the first place, nicad batteries are safer. There is no flammable hydrogen gas vented to the atmosphere, and no caustic sulfuric acid. In contrast to the lead—acid cell's life of less than 400 charge—discharge cycles, nicad cells can be recharged thousands of times. Other advantages include compactness and low weight combined with terrific power outputs. The major drawback is charge leakage. These cells must be periodically recharged whether used or not.

The arrangement in Fig. 7-43 is typical. The battery pack is mounted on the handle and the charger plugs into a 110V outlet. A full charge requires between 16 hours at one-tenth of the ampere-hour rating. For example, a 20 A-hr nicad should be charged with a 2A input current for 16 hours. Trickle-charging is not critical. The batteries can be

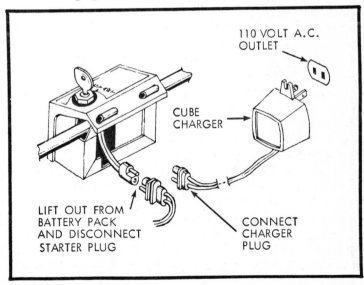

110 VOLT A.C. OUTLET

CUBE CHARGER

LIFT OUT FROM BATTERY PACK AND DISCONNECT STARTER PLUG

CONNECT CHARGER PLUG

Fig. 7-43 Nicad battery pack and charger (Courtesy Ford Tractor Operations)

recharged as long as temperatures do not drop below 40°F or climb above 105°F. Charging is not harmful if limited to 7 to 10 days.

The starter motors used with nicad systems are 12V types, similar to the ones which have already been discussed.

The battery pack can be checked by loading it with a pair of GE 4001 headlamps connected in parallel as in Fig. 7-44. The exact capacity varies with the battery-pack manufacturer. Although there will be differences between brands, the bulbs should burn for 5 minutes with a freshly charged pack. Briggs

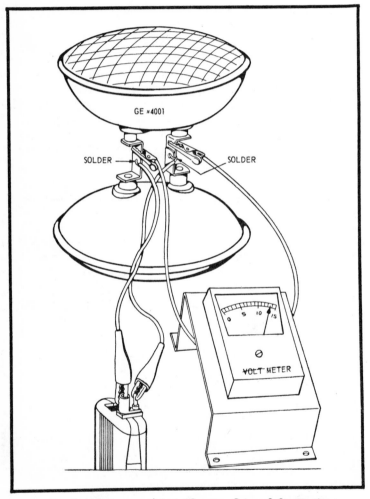

Fig. 7-44 Battery-pack test (Courtesy Briggs & Stratton)

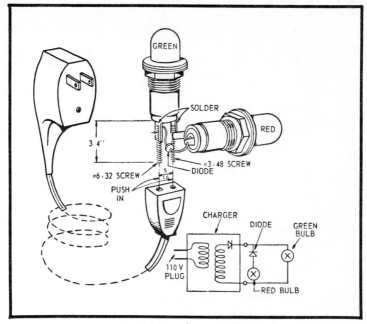

Fig. 7-45 Briggs charger tester.

goes further and specifies that the voltage should not fall below 13.5V after 1 minute. A 13V reading is hard evidence that the pack should be replaced.

Each manufacturer uses his own charger—a fact which makes evaluation of output difficult. Tecumseh supplies a tester (part No. 670235) for use with their chargers. It will be inaccurate with other makes and may damage them as well. Briggs suggests that the technician build his own. Parts required are:

No. 53 bulbs (2)
Lamp socket—Dialco No. 0931-102 (red)
Lamp socket—Dialco No. 0932-102 (green)
1N5061 diode
No. 6-32, ¾ in. machine screw
No. 3-48, ¾ in. machine screw

These parts are assembled as shown in Fig. 7-45. To use the tester, plug it into the charger. If the green lamp comes on, the tester is good. If both lamps light, the tester's diode is shorted. Failure of either to light means an open transformer winding or diode.

Chapter 8

Major Engine Work

Although Briggs & Stratton is rumored to be working on a "throw-away" engine, the days of cheap power mowers seem to be over. Customers have become more sophisticated and demanding in their tastes. The government has intervened to make mowers safer, quieter, and generally more civilized. As a result, the $30 lawnmower is an anachronism. Major engine repairs which involve several days of concentrated labor and high parts costs are becoming more routine. Just a few years ago, boring cylinders was almost unheard of. Mechanics in large shops vied for these jobs to relieve boredom. Today there is nothing unusual about it. Customers may demand two or more overbores on the same block and sometimes go through several crankshafts in the process.

For convenience, major engine repairs are divided into three categories. An *overhaul* means a valve and ring job. In some shops you will get just that—nothing more. Other shops take responsibility for *all* systems during an overhaul. They clean and rebuild the carburetor, replace the wearing ignition components, drive belts, and starter cords. In this context, an overhaul is more of a renovation, and may involve 2 full working days. A *rebuild* means that every critical engine part is inspected and refurbished as needed. It is assumed that the cylinder (cast iron) will be bored; but replacement of the rod and main bearings could be included. Again, the larger and more prosperous shops check out all systems and do required repairs. A thorough rebuild is expensive, but the end product is within striking distance of a new machine. The alternative is to install a *short block*. These assemblies are available by special order from the distributor. They consist of a new block, camshaft, valves (side-valve engines), piston, rod, crankshaft, and flange (Fig. 8-1), assembled and inspected by the factory.

All the mechanic need do is install the head, magneto, carburetor, and other engine accessories. A short block is practically the equivalent of a new engine, and has all of the reliability and durability associated with new parts and factory inspection norms. Short blocks cost between 50 and 60% of the price of a new engine. The picture is clouded by the fact that blocks are not available for all models; for some, the crankshafts have to be changed. Kohler supplies a *mini block* which is equivalent to a short block, less crankshaft.

PRETEARDOWN INSPECTION

Before lifting a wrench to the engine, make as complete an inspection as you can. The British Army adage (to the effect) that time spent in reconnaissance is not wasted also applies to major engine repairs—as well as to warfare. An inspection may convince you to drop the whole project and install a new engine or, if the damage extends outside of the engine, to purchase a new mower. Careful inspection can save you the time and effort of extraneous teardowns. Unless there is strong evidence to the contrary (as when an engine has been run without oil), complete and total disassembly is to be avoided. Engines are made to be taken apart, but they do not take kindly to it.

Begin your assessment with a compression test. The typical lawnmower engine should develop between 60 and 80 psi during cranking. In general, 2-cycle engines have lower compression ratios than 4-cycles and, consequently, develop lower cranking pressures. At any rate, you can consider 45 psi to be the absolute minimum pressure allowable. Below 40 psi, it is almost impossible to start the engine. (Make the test as detailed in Chapter 3.) Note: Engines with compression-release devices must be turned backwards to obtain true readings.

Another test for Kohler and other negative-pressure sump engines, involves the use of a manometer. While this test is rarely used, it is highly instructive—the effort to build one is worthwhile. Procure a length of clear plastic tubing (such as a Neoprene fuel line) and mount it on a board as shown in Fig. 8-2. Make a gentle bend in the tube. Now, mark off a foot in inch increments on the board. Next, fit a rubber stopper snugly in the oil filler hole. Drill a hole in the stopper to accept the end of the tube. Fill the tube with water (vegetable coloring makes it easier to read) until the fluid level reaches the halfway mark

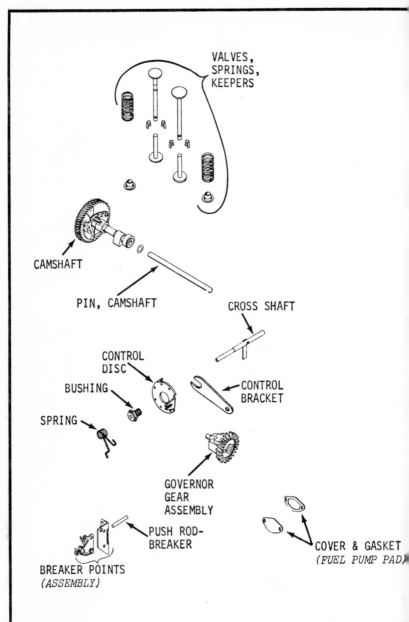

VALVES, SPRINGS, KEEPERS

CAMSHAFT

PIN, CAMSHAFT

CROSS SHAFT

CONTROL DISC

BUSHING

CONTROL BRACKET

SPRING

GOVERNOR GEAR ASSEMBLY

PUSH ROD-BREAKER

COVER & GASKET *(FUEL PUMP PAD)*

BREAKER POINTS *(ASSEMBLY)*

NOTE: All parts are assembled to short block except oil fill tube and dipstick.

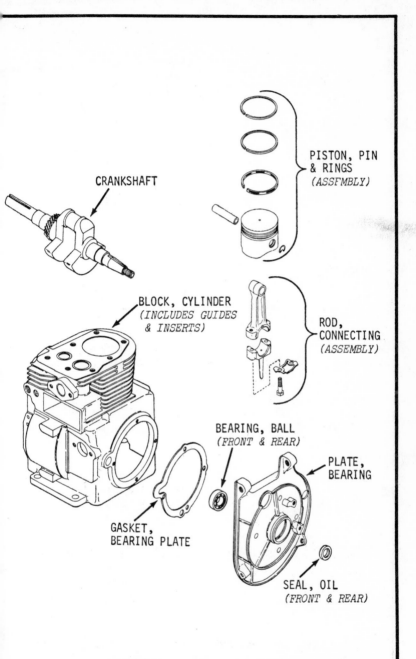

CRANKSHAFT

PISTON, PIN
& RINGS
(ASSFMBLY)

BLOCK, CYLINDER
*(INCLUDES GUIDES
& INSERTS)*

ROD,
CONNECTING
(ASSEMBLY)

BEARING, BALL
(FRONT & REAR)

PLATE,
BEARING

GASKET,
BEARING PLATE

SEAL, OIL
(FRONT & REAR)

Fig. 8-1 Short block assembly (Courtesy Kohler of Kohler)

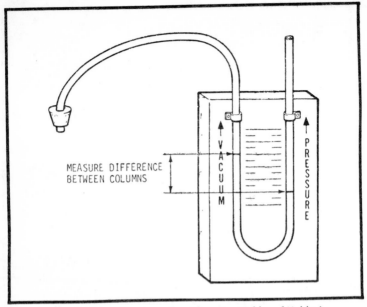

Fig. 8-2 U-tube manometer (Courtesy Kohler of Kohler)

on the scale. Plug the manometer into the engine and measure the difference in fluid levels. The crankcase should be under partial vacuum, causing the instrument to register between 5 and 10 in. on the near side of the tube. A vacuum gage will show 0.5 to 1 in. of mercury (12.5−25 torr). No vacuum is a sign of leaking oil seals, a pinhole in the case, or a leaking flange gasket. Positive pressure will develop if the oil breather is clogged or if the rings and valves leak.

Tests for bearing wear and the various knocks and rattles associated with impending mechanical doom are discussed in Chapter 3.

Compression and crankcase vacuum tests, coupled with a close visual inspection, will give you a good indication of the repairs needed. The troubleshooting guides (Figs. 8-3 and 8-4) are quite comprehensive. Figure 8-3 can be applied to 2-strokes as well—if you keep in mind the absence of valve gear in those machines, and account for exhaust port and muffler clogging.

GENERAL REPAIR PROCEDURE

There are certain rules proved valid by experience with large, complex assemblies. To begin with, all parts must be

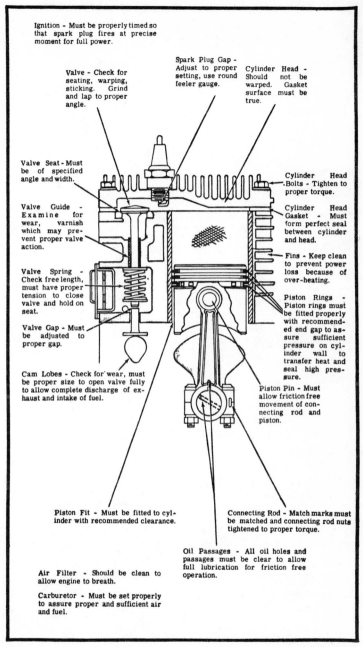

Ignition - Must be properly timed so that spark plug fires at precise moment for full power.

Spark Plug Gap - Adjust to proper setting, use round feeler gauge.

Cylinder Head - Should not be warped. Gasket surface must be true.

Valve - Check for seating, warping, sticking. Grind and lap to proper angle.

Valve Seat - Must be of specified angle and width.

Cylinder Head Bolts - Tighten to proper torque.

Valve Guide - Examine for wear, varnish which may prevent proper valve action.

Cylinder Head Gasket - Must form perfect seal between cylinder and head.

Fins - Keep clean to prevent power loss because of over-heating.

Valve Spring - Check free length, must have proper tension to close valve and hold on seat.

Piston Rings - Piston rings must be fitted properly with recommended end gap to assure sufficient pressure on cylinder wall to transfer heat and seal high pressure.

Valve Gap - Must be adjusted to proper gap.

Cam Lobes - Check for wear, must be proper size to open valve fully to allow complete discharge of exhaust and intake of fuel.

Piston Pin - Must allow friction free movement of connecting rod and piston.

Piston Fit - Must be fitted to cylinder with recommended clearance.

Connecting Rod - Match marks must be matched and connecting rod nuts tightened to proper torque.

Oil Passages - All oil holes and passages must be clear to allow full lubrication for friction free operation.

Air Filter - Should be clean to allow engine to breath.

Carburetor - Must be set properly to assure proper and sufficient air and fuel.

Fig. 8-3 Points to check for engine power (Courtesy Tecumseh Products Co.)

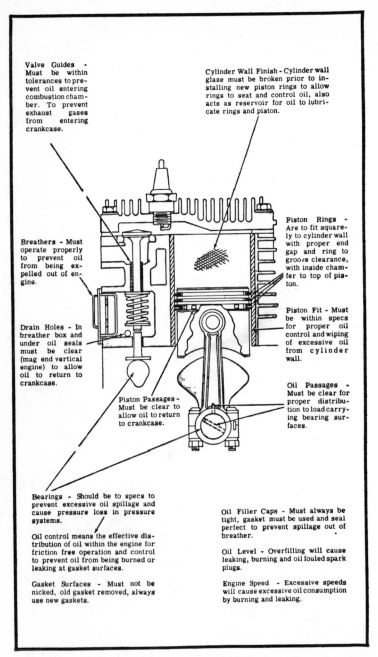

Valve Guides - Must be within tolerances to prevent oil entering combustion chamber. To prevent exhaust gases from entering crankcase.

Cylinder Wall Finish - Cylinder wall glaze must be broken prior to installing new piston rings to allow rings to seat and control oil, also acts as reservoir for oil to lubricate rings and piston.

Piston Rings - Are to fit squarely to cylinder wall with proper end gap and ring to groove clearance, with inside chamfer to top of piston.

Breathers - Must operate properly to prevent oil from being expelled out of engine.

Piston Fit - Must be within specs for proper oil control and wiping of excessive oil from cylinder wall.

Drain Holes - In breather box and under oil seals must be clear (mag end vertical engine) to allow oil to return to crankcase.

Oil Passages - Must be clear for proper distribution to load carrying bearing surfaces.

Piston Passages - Must be clear to allow oil to return to crankcase.

Bearings - Should be to specs to prevent excessive oil spillage and cause pressure loss in pressure systems.

Oil control means the effective distribution of oil within the engine for friction free operation and control to prevent oil from being burned or leaking at gasket surfaces.

Gasket Surfaces - Must not be nicked, old gasket removed, always use new gaskets.

Oil Filler Caps - Must always be tight, gasket must be used and seal perfect to prevent spillage out of breather.

Oil Level - Overfilling will cause leaking, burning and oil fouled spark plugs.

Engine Speed - Excessive speeds will cause excessive oil consumption by burning and leaking.

Fig. 8-4 Points to check for engine oil control—4-cycles (Courtesy Tecumseh Products Co.)

Bolt or Stud Diameter	Type 1 Studs Only		Type 1 Bolts 6" length or less		Type 1 Bolts longer than 6"		Type 5 (all lengths)		Type 8 (all lengths)			
									Only when used† in cast (gray) iron		All other applications	
	Min.	Max	Min	Max	Min	Max	Min	Max	Min.	Max.	Min.	Max.
1/4	5	6	5	6	3	3	9	10	11	13	12	14
5/16	12	13	12	13	6	7	19	21	24	27	27	30
3/8	21	24	21	24	11	13	33	37	43	47	45	50
7/16	35	38	35	38	19	21	53	60	69	76	75	85
1/2	52	58	52	58	29	32	80	90	104	117	115	130
9/16	70	80	70	80	41	46	115	130	150	170	165	185
5/8	98	110	98	110	57	63	160	180	210	230	220	250
3/4	174	195	174	195	100	112	290	320	350	390	400	450
7/8	300	330	162	181	162	181	420	470	570	630	650	730
1	420	470	250	270	250	270	630	710	850	950	970	1090
1-1/8	600	660	350	380	350	380	850	950	1200	1350	1380	1550
1-1/4	840	940	490	540	490	540	1200	1350	1700	1900	1940	2180
1-3/8	1100	1230	640	710	640	710	1570	1760	2300	2500	2600	2800
1-1/2	1470	1640	850	940	850	940	2000	2300	3000	3300	3300	3700
1-3/4	2350	2450	1330	1490	1330	1490	3300	3700	4700	5200	5300	6000
2	3500	3900	2000	2200	2000	2200	5000	5500	7000	7800	8000	9000

† When bolt penetration is 1-1/2 times the diameter of the bolt.

NOTE: Multiply the standard torque by:

.65 when finished jam nuts are used.
.70 when Molykote, white lead or similar mixtures are used as lubricants.
.75 when parkerized bolts or nuts are used.
.85 when cadmium plated bolts or nuts and zinc bolts w/waxed zinc nuts are used.
.90 when hardened surfaces are used under the nut or bolt head.

Fig. 8-5A Torque specifications—nongasketed joints only (Courtesy International Harvester)

cleaned. The external surfaces of the engine should be washed down with Gunk and hosed off prior to disassembly. Scrub each part with kerosene as it is removed. Cleaning will reduce bearing damage on startup, and will insure that the parts have been allowed adequate visual inspection.

An engine is a collection of bearing surfaces; these surfaces have to be finished smoothly, and must n.eet clearance specifications. Ideally, all moving parts should be miked (measured with a micrometer), but this is rarely done, since mechanics pride themselves on their vernier eyeballs.

All critical parts—head, flange, rod journal, flywheel nut, and crankcase halves—should be torqued upon assembly. The torque chart in Fig. 8-5A is offered as a general guide, courtesy of International Harvester. The specifications given do not apply to gasketed joints. (For other torque specs, see Tables 8-1 and 8-2.)

Lubricate all moving parts immediately prior to assembly. For the first few revolutions the only lubrication the parts have is that which you provide. A pumper oil can is

IH TYPE	S.A.E. GRADE	DESCRIPTION	BOLT HEAD MARKING *
1	(equivalent) 1 or 2	WILL HAVE A ᵕᴾ STANDARD MONOGRAM IN THE CENTER OF THE HEAD Low or Medium Carbon Steel Not Heat Treated	 (ᵕᴾ)
5	5	WILL HAVE A ᵕᴾ AND 3 RADIAL LINES Quenched and Tempered Medium Carbon Steel	 (ᵕᴾ)
8	8	WILL HAVE A ᵕᴾ AND 6 RADIAL LINES Quenched and Tempered Special Carbon or Alloy Steel	 (ᵕᴾ)

*The center marking identifies the bolt manufacturer. The ᵕᴾ monogram is currently used.
Some bolts may still have an IH or a raised dot which previously identified IH bolts.

Fig. 8-5B Bolt identification chart (Courtesy International Harvester)

convenient for this. Use high-grade motor oil of the type specified for the crankcase, but do not dilute it with additives. Replace all gaskets with new ones to insure that the engine will be oil-tight. It is good practice to replace the oil seals during an overhaul—and mandatory in a rebuild.

In general, the work will go better if you specify factory-original parts. If you're on a tight budget, there are a number of small-engine "bootleg" parts around whose quality is indeterminable; however, you can purchase factory-equivalent antifriction bearings. Continental makes excellent replacement piston rings for small engines. The chart in Fig. 8-5B gives standard fastener identification codes. If you plan to paint the block after assembly, use the original color (available from the distributor in spray cans) but avoid metallic-base paints. An engine stand (Fig. 8-6) always makes this sort of work more convenient; two models are available from Clinton Engines Corporation. The plans for a stand used with vertical crankshaft engines are given in Fig. 8-6.

Cylinder Heads

Cylinder heads can be removed with the engine in place on the machine. It is generally easier to dismantle the hood paneling (on riding mowers and tractors), than it is to pull the engine. After removing the engine shroud and the spark plug, note the length of the cylinder head bolts. Briggs & Stratton, for example, uses bolts that are longer in the exhaust valve area.

Carbon deposits can be cleaned from the head and block (Fig. 8-7) with a wire scraper or a dulled cutting tool and finish with a power-driven wire wheel. Do not scratch the gasket surfaces or gouge the chamber walls; scratches and nicks can

Table 8-1. Rod and Head Torquing Specifications.

	Connecting Rod	Cylinder Head
Briggs & Stratton 6B, 6000, 8B, 80000, 82000, 92000, 100000, 130000	100 in.-lb	140 in.-lb
140000, 170000, 171700, 190000, 191700	165 in.-lb	165 in.-lb
5, 6, N, 8	100 in.-lb	140 in.-lb
9	140 in.-lb	140 in.-lb
19, 190000, 200000, 23, 230000, 243000, 300000, 320000	190 in.-lb	190 in.-lb
Clinton Gem, Clintalloy Long Life Panther GK 590	70 – 80 in.-lb 70 – 80 in.-lb 70 – 80 in.-lb 70 – 80 in.-lb	125 – 150 in.-lb 200 – 220 in.-lb not available not available
200, A200, AVS200, VS200, VS400, AVS400, BVS400, CVS400, A400, A490	35 – 45 in.-lb	not available
Red Horse (all)	130 – 140 in.-lb	200 – 220 in.-lb
J200, J300	35 – 40 in.-lb	not available
J350, J500, J700	70 – 80 in.-lb (forged aluminum)	not available
Kohler K91 K141, K161, K181	140 in.-lb 200 in.-lb	200 in.-lb 15 – 20 ft.-lb
K242, K301, K321, K341 K482, K532 K662	300 in.-lb 25 ft.-lb 35 ft.-lb	25 – 30 ft.-lb 35 ft-lb 40 ft.-lb
Outboard Marine Corp. A, C, D	55 – 60 in.-lb	100 – 125 in.-lb (cylinder to crankcase)
Tecumseh Products Lauson 1.5 – 3.5 hp	65 – 75 in.-lb	140 – 200 in.-lb
4 – 5 hp Small Frame (LAV50, HS 50, LAV40, HS40)	80 – 95 in.-lb	140 – 200 in.-lb
4 – 6 hp Medium Frame (includes H40, HH40, V40, VH40, V50, VH50, H50, HH50	86 – 110 in.-lb	140 – 200 in.-lb
7 – 8 hp Medium Frame 8 hp and over—Heavy Frame	106 – 130 in.-lb 86 – 110 in.-lb	140 – 200 in.-lb 200 – in.-lb

cause leaks and will make carbon removal more difficult the
next time around. Caution: If the gasket has stuck and become
torn, scrape the remnants off with a knife—do not use a wire
brush on this material. Head and exhaust gaskets (as well as
intake gaskets on Lauson engines) are made of asbestos. The
dust that results from brushing that material is extremely
hazardous to breathe.

Table 8-2. Fastener, and Torque Specifications.

aluminum and bronze rods	10—24 × ¹⁷/₃₂ in. Filister screws	40—50 in. —lb upon assembly, 30—40 in. —lb after 1 hour running
steel rod to rod cap	10—32 × ⁹/₁₆ in. socket head screws	70—80 in. —lb
steel rod to rod cap	10—32 × ⅝ in. socket head screws	70—80 in. lb
cylinder to crankcase nuts	¼ in. —20 hex head screws	70—75 in. —lb on assembly, 20 in. —lb after one hour running
cylinder head to cylinder	10—24 × ¾ in. pan head machines screws	30—40 in. —lb
cylinder head to cylinder	10—24 × ¾ in. socket head cap screws	50—60 in. —lb

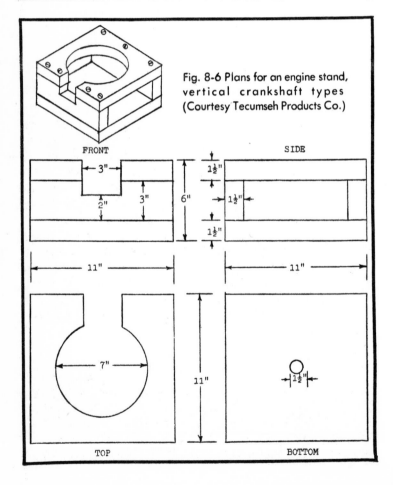

Fig. 8-6 Plans for an engine stand, vertical crankshaft types (Courtesy Tecumseh Products Co.)

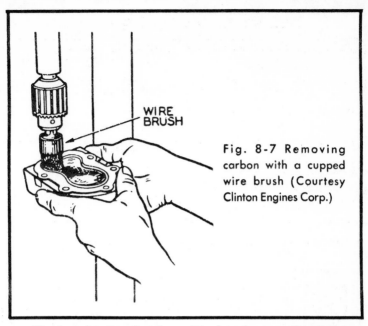

WIRE BRUSH

Fig. 8-7 Removing carbon with a cupped wire brush (Courtesy Clinton Engines Corp.)

The bench setup for the critical endeavor of grinding a head requires that you tape a piece of No. 340 (wet or dry) sandpaper on plate glass, a drill-press table, or other flat surface. Using motor oil as a lubricant, grind the head until all parts of the gasket surface are shiny. Hold the head in the center and use a figure-8 motion for even cutting. If the head is badly warped, you can speed the process with No. 280, or harsher, grit paper. Graduate to 9 fine grit for finishing.

The spark plug boss should be scrutinized for stripped or otherwise damaged threads. You can renew threads with a 14 mm thread chaser. Stripped threads mean that the head should be replaced, although it is usually cheaper to have a shop install a specially tapped insert. Many lawnmower, motorcycle, and outboard repair stations have the necessary inserts and reamers on hand.

Valves

To remove the valves from side-valve engines, remove the breather first. Insert a valve-spring compressor according to Fig. 8-8. The retainer pictured is a type common to small Briggs & Stratton engines. On early models you will find the valve secured by a pin. Other Briggs engines and the products of most other manufacturers employ the split collets shown, or

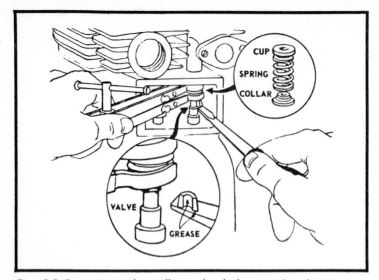

Fig. 8-8 Removing and installing valve locks on side-valve engines (Courtesy Briggs & Stratton)

ones like those in Fig. 8-9. Be careful; these parts could fall into the crankcase through the valve chamber port. Segregate the springs and the keepers to prevent a mixup during assembly.

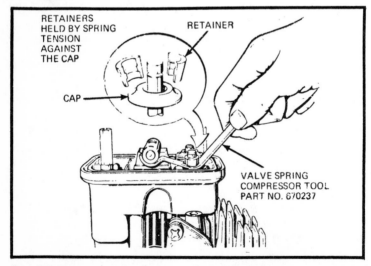

Fig. 8-9 Removing valve locks on the Lauson HH 150 OHV engine (Courtesy Tecumseh Products Co.)

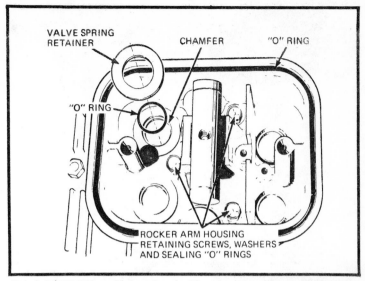

Fig. 8-10 HH 150 gaskets and seals (Courtesy Tecumseh Products Co.)

The valves in Lauson OHV engines are removed between the stem and the rocker shaft (Fig. 8-9) with a compressor bar. Note the *O*-rings (Fig. 8-10) below the valve spring retainer washer that are around the rim of the housing, and under the rocker arm retaining screws. Replace these rings as you would any other gasket. Look for wear and scoring on the

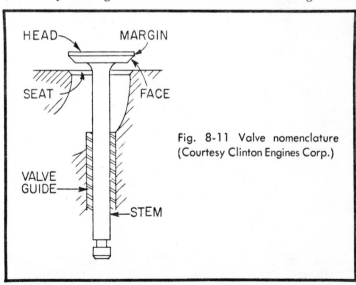

Fig. 8-11 Valve nomenclature (Courtesy Clinton Engines Corp.)

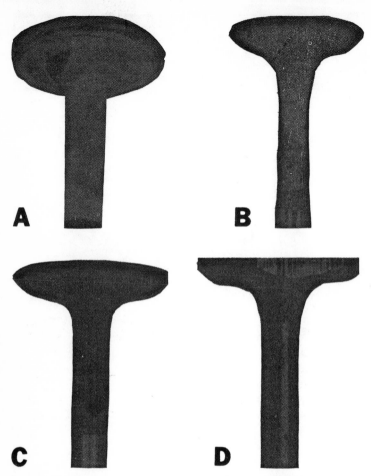

Fig. 8-12 Exhaust valve: (A) normal; (B) stem corrosion; (C) overheating; (D) carbon cut. (Courtesy Kohler of Kohler)

rocker arm shaft. It is integral with the rocker arm housing and therefore not a replaceable part.

After long service the exhaust valve (Fig. 8-11) and stem will accumulate granular deposits (Fig. 8-12A) which may be brown, yellowish, or tan. This condition is perfectly normal. Stem corrosion (Fig. 8-12B) is not; it is caused by moisture in the fuel or from condensation. It also occurs when the engine is repeatedly stopped before it reaches operating temperature. If your inspection reveals this to be the case, replace the valve. Overheating of the exhaust valve is shown by a black, polished

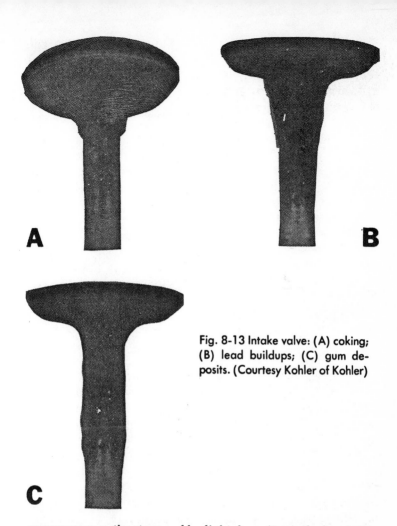

Fig. 8-13 Intake valve: (A) coking; (B) lead buildups; (C) gum deposits. (Courtesy Kohler of Kohler)

appearance on the stem and by light deposits on the head (Fig. 8-12C). Worn guides and weak springs may be the cause, along with lean carburetor settings. Carbon cutting occurs infrequently and usually is associated with engines which pump oil. The carbon becomes so hard that it slices the valve as it did to the one in Fig. 8-12D.

Since the intake valve is cooled by the incoming charge, it encounters less severe problems than its brother on the exhaust side. Coking, or heavy carbon formations under the head (Fig. 8-13A) is normal and, assuming the valve is otherwise okay, it can be cleaned and put back into service. Lead buildup (Fig. 8-13B) is another sign of normal wear.

Some leakage around the seat is evident in the photo. This condition is distinguished by the formation of deposits; carbon cakes in layers and lead drops out of the fuel in globules. Gum and varnish deposits (Fig. 8-13C) are caused by stale gasoline or by shutdown before the engine has warmed. Expect the rings to be affected as well.

Another symptom of poor valve condition is pronounced wear on the stem. The surest way to determine this is to mike the stem against factory specifications. Margin dimensions (Fig. 8-14) should also be verified; sharp edges will cause the valve to overheat and may bring on preignition. A bend in the stem can easily be seen as a wobble when the valve is chucked in a drill motor, or press, and spun. Replace the valve—it is almost impossible to get a good seal with this condition. Be particular about worn keeper grooves on overhead valve engines. A swallowed valve is catastrophic.

Light-metal (and most cast-iron) block engines have removable valve seat inserts secured by an interference fit. Seats fail by erosion (in which case they can be reground), by cracking, or by coming adrift from the block. Figure 8-15 shows a simple removal tool which can be ordered from Briggs & Stratton or fabricated in the shop. Clinton inserts have a metal lip rolled over the outer circumference to help hold them in place. After the lip has been cut away, the inserts are driven out from behind with a long punch (Fig. 8-16). The recess is cut to a depth equal to insert size plus $1/32$ in. for peening. Use Clinton tool 95115; other manufacturers supply their own reamers, but the installation procedure follows the same outline. Clean the bore with compressed air so that no metal chips remain. By cleaning before the cutter pilot is removed from the valve guide, you assure that chips do not get into the valve chamber. The same reaming process can be used to fit

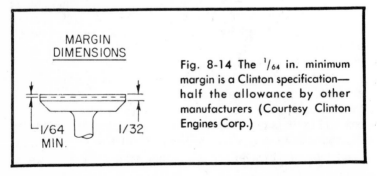

MARGIN
DIMENSIONS

1/64
MIN.

1/32

Fig. 8-14 The $1/64$ in. minimum margin is a Clinton specification— half the allowance by other manufacturers (Courtesy Clinton Engines Corp.)

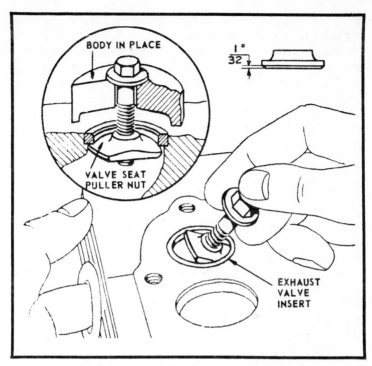

Fig. 8-15 Valve seat removal tool (Courtesy Briggs & Stratton)

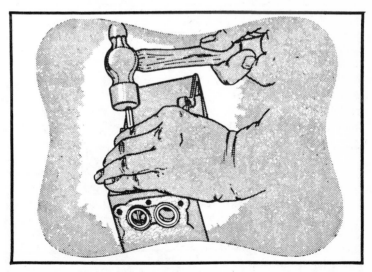

Fig. 8-16 Removing Clinton valve seats (Courtesy Clinton Engines Corp.)

inserts in cast-iron blocks which were not originally equipped with them.

Typically, the replacement insert has an outside diameter .0020 in. greater than the inside diameter of the recess. Chill the insert with frozen CO_2 (dry ice) and warm the block with a propane torch before installation with a factory driver (most shops use an old valve). Work quickly while the parts are still at different temperatures. Finish by peening over the seat. The Clinton arrangement is illustrated in Fig. 8-18. Briggs & Stratton seat inserts, as well as other makes, are stacked flush (Fig. 8-19).

Stacking is an art, especially when one is dealing with a light-metal block. Begin by crisscrossing to hold the insert in place. Some upper cylinder bore distortion is inevitable in the process. The amount of distortion depends on the tool's angle off vertical, and the impact used. Staking should be done before the cylinder is resized or deglazed. Aluminum engines should be run for 2 hours at no or half load. The heat generated in operation will relieve the stresses and minimize distortion. Another method is to heat the block in an oven for several hours at 350°F. (Allow it to cool down slowly.)

Grinding, or lapping, valves is almost outmoded, but has some utility. The process can remove small imperfections.

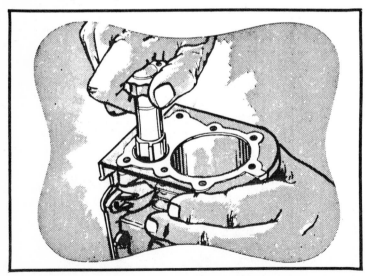

Fig. 8-17 Reaming the recess. Reamers and pilots vary between engine makes and models (Courtesy Clinton Engines Corp.)

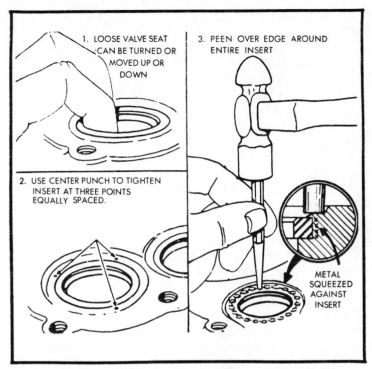

Fig. 8-18 Seat insert chamfer (Courtesy Clinton Engines Corp.)

Heavy grinding is self-defeating, since the block will have grown under heat. Figure 8-20A shows a perfectly lapped seat with the engine cold; the drawing at B shows the same seat when the engine reaches operating temperature. This

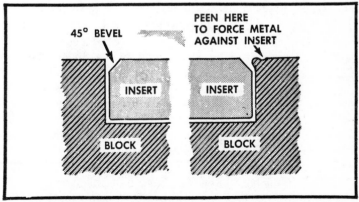

Fig. 8-19 Flat-topped insert installation (Courtesy Briggs & Stratton)

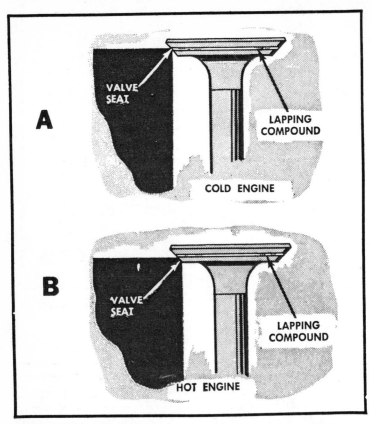

Fig. 8-20 (A and B) The effects of heat on a heavily lapped valve and seat (Courtesy Clinton Engines Corp.)

phenomenon is present in all engines and is most pronounced in aluminum-block types. Unless you are working on an antique like the Briggs Model N, you will need a suction cup to hold the valve (Fig. 8-21). Older engines had indentations in the valve face to accept a screwdriver bit or other tool. K-D makes a small engine suction cup as part No. 501. Because the area involved is small, the cups keep coming unstuck. One way to hold the cup in place is to make the seal airtight with gasket varnish. Another solution, which came to me after much slipping and sliding, is to secure a dowel pin to the valve head with a tiny amount of alpha cyanoacrylic contact cement. You can dissolve the bond later with fingernail polish remover, or merely break the pin. A few grains of cellulose adhering to the valve does not seem to hurt anything.

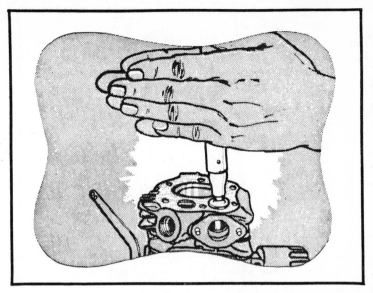

Fig. 8-21 Lapping valves with a suction cup (Courtesy Clinton Engines Corp.)

Lapping compound comes in small, double-walled tins. One half is coarse for rapid (a relative term) material removal, and the other is fine for finish lapping. Put a small amount of compound on the valve face and rotate the suction cup between your palms in a manner reminiscent of a Boy Scout making a fire. The valve will make a dull swishing noise as the compound cuts. Periodically, raise the valve slightly to allow the compound to run down on the face, and rotate it so that the whole circumference is progressively lapped. Replenish the compound as needed. Running the valve dry will make deep scratches which are almost impossible to lap out.

There are several ways to tell when you are finished. One is to wipe the compound off and drop the valve. It should make a dull clunk—like a bogus fifty-cent piece. If the valve rings, you know that all surfaces are not in contact. Another method is to paint the seat with Prussian Blue and give the valve a half-twist. The Blue should be broken all the way around the seat. Perfectionists coat the valve with liquid soap, or air-conditioning leak detector, and force air through the ports. You should be able to stop a valve from leaking with thumb pressure.

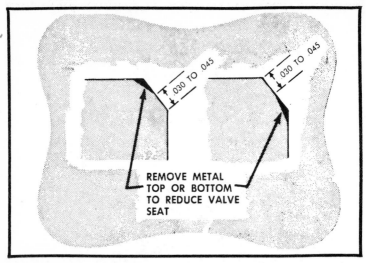

Fig. 8-22 Valve seat width—critical for best performance (Courtesy Clinton Engines Corp.)

If the valves do not respond to lapping, they must be machine ground. The limits of lapping are shown by persistent leaks, visible loss of valve metal, overly wide seats, or scratches. The latter may be due to loss of compound or, more commonly, broad seats. Small engine seat widths vary between make and model, but $^1/_{32}$ to $^3/_{64}$ in. is a good average.

The seats are narrowed with a cutting tool at either flank as shown in Fig. 8-22. The seat angle is generally 46° to mate with a 45° valve angle. Clinton specifications make the seats with a more acute angle of 43½° to 44½°. Many older engines had 30° seats. Briggs intakes are still made at this reduced angle.

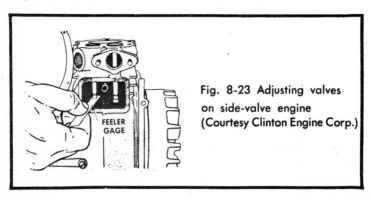

Fig. 8-23 Adjusting valves on side-valve engine (Courtesy Clinton Engine Corp.)

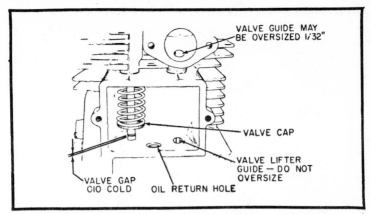

Fig. 8-24 Side-valve lash—use a feeler gage between the tappet and valve spring, cam lobe down (Courtesy Tecumseh Products Co.)

Both manual and power-driven valve seat grinders are available. Most shops invest in a powered type, or farm the work out to an automotive machinist.

After the valves are ground or lapped, the lash (clearance) must be set. Side-valve engines do not have a provision for this adjustment; the end of the valve stem must be ground. It should be dead flat and parallel with the head, which means that you need to cradle the valve in **V**-blocks as you hold it against the wheel. Unless you have a very light touch, finish with a file. Go slowly, since it is easy to take off too much metal. Good average clearance figures for most small engines are 0.001 in., measured with a feeler gage as illustrated in Fig. 8-24.

Lash adjustment for the Lauson OHV is much simpler. Loosen the locknuts, turn the adjusting screws, and insert a feeler gage between the stem and the rocker arm (Fig. 8-25). Holding the locknuts with a wrench, turn them clockwise to decrease the lash, or counterclockwise to increase it (Fig. 8-26).

Valve guides have three functions; they hold the valve concentric to the insert, seal air out of the combustion chamber, and act as heatsinks to cool the valve. Surprisingly, many light-frame blocks do not have guides as such; the valves run directly on the aluminum-alloy block material. The manufacturers who have taken this shortcut suggest two remedies; either the bores can be reamed oversize and larger-stemmed valves fitted (the usual oversize is $1/32$ in.) or

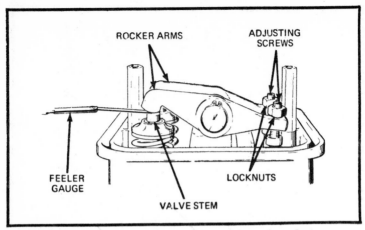

Fig. 8-25 OHV lash (Courtesy Tecumseh Products Co.)

the bores can be cut to accept valve guides stocked for other engines. The second choice is the best solution, but may take some rummaging amid the parts stores to find acceptable guides.

Existing guides can be knurled oversize with a tool from Clinton. Knurling also works on tappet guides which are integral with the block. Should knurling be out of the question—either because the tool is not available or because

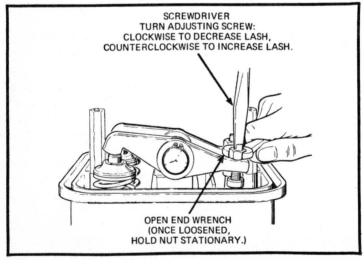

Fig. 8-26 Adjusting OHV lash—0.005 in. intake and 0.010 in. exhaust, cold (Courtesy Tecumseh Products Co.)

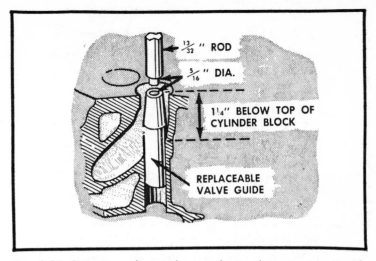

Fig. 8-27 Removing valve guide—punch tip dimensions vary with application (Courtesy Clinton Engines Corp.)

wear has exceeded 0.004 in.—replace the guide. Figure 8-27 shows an appropriate punch. The tip dimension should be the diameter of the valve stem. New guides are chilled and pressed into a heated block from above. Finish reaming is not necessary if the temperature differential was such that the guide moved into place without deformation.

Pistons

The piston may be removed from the sump- or flange-side on 4-cycle engines. The crankshaft must be free of rust and dirt accumulations before the flange is parted. These accumulations become serious on rotary lawnmowers, where the shaft is exposed to rocks and pebbles under the deck. Remove the blade and blade adapter. You may have to use a puller on the adapter if it has rusted. Another means of persuasion is to shock the adapter loose with a pair of machinist's hammers. Strike both sides of the adapter hub simultaneously to drive it down on the shaft. Turn the shaft periodically to distribute the blows. Penetrating oil may help, especially if you allow a few days for it to soak into the joint.

Once the adapter is free, remove the half-moon keys from the shaft with a pair of side-cutting pliers. Chances are the shaft metal will have billowed a few thousandths of an inch around the keys. File this material away, being careful not to nick the shaft. Remove all rust with strip emery cloth. Take

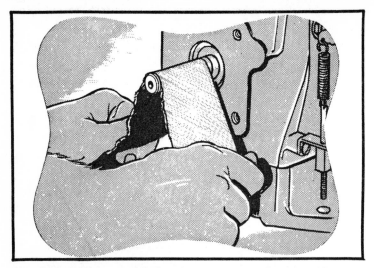

Fig. 8-28 Polishing crankshaft prior to opening the case (Courtesy Clinton Engine Corp.)

your time—any imperfections left on the shaft will scratch the lower main bearing as the flange comes off (Fig. 8-28). Light taps with a mallet will release the gasket and alignment pins. Imperfections on shafts which are supported by antifriction bearings will lock the shaft to the inner race. Pistons are accessible in 2-cycle engines by lifting the cylinder barrel.

Before disassembly, observe the lay of the piston and rod, and look for oil ports, casting numbers, and other points of

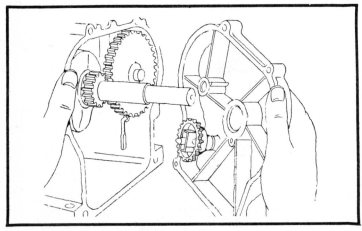

Fig. 8-29 Flange removal (Courtesy International Harvester)

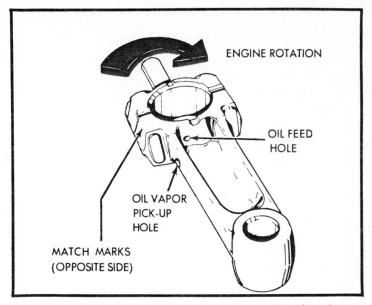

Fig. 8-30 Rod reference marks (Courtesy Tecumseh Products Co.)

reference. In many engines the piston is offset and must be assembled relative to the rod in only one way; otherwise, the engine will knock. If you do not find marks or other nonsymmetrical features, scribe your own on the inside of the piston and on the rod shank. Figure 8-30 shows the lay of rods for VH80 and VH100 Lauson engines. The picture is not an accurate guide in so far as other engine types and makes are concerned.

All connecting rods have match marks between the big-end journal halves. These match marks *must* be aligned upon assembly; the marks are usually embossed (Fig. 8-31). However, they may consist of a pair of scribe marks or punch pricks.

The piston wristpin may have a definite left or right orientation as defined by the flywheel and power-takeoff sides of the engine. Briggs pins are installed with the solid end toward the match marks. In any case, the pin should be replaced as it was originally installed. Small engines employ *full-floating* pins; that is, the pin "floats" on both of the piston bosses and on the small end of the rod. At room temperatures the piston shrinks so much that the pin must be forced out. Remove the two circlips with long-nose pliers and discard

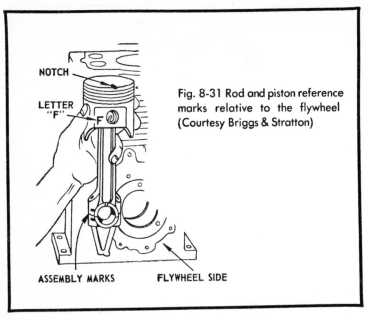

NOTCH

LETTER "F"

ASSEMBLY MARKS

FLYWHEEL SIDE

Fig. 8-31 Rod and piston reference marks relative to the flywheel (Courtesy Briggs & Stratton)

them. Circlips cost about a dime apiece and are good insurance. Ideally, the pin should be removed or installed with the tool shown in Fig. 8-32. Pin presses are available from Kohler or from some motorcycle dealers. If you have an acetylene torch, you can easily make one up using a $^7/_{16}$ in. bolt and nut. Otherwise, drive the pin out with a punch, being very careful not to bend the rod or scratch the piston bosses. Drive it far enough so that the small end of the rod clears the pin.

The job is slightly complicated for the better 2-cycles that have uncaged needle bearings in the small end. Keep track of the bearings and note the position of the thrust washers on either side of the rod.

Inspect the pin for wear; the piston pin bosses should be free from scoring or cracks. Small end fit should be loose enough to allow the piston to fall to the right or left by its own weight. But there should be no perceptible up-and-down play.

Remove the rings with an expander if possible. You can remove them by hand—but rings are razor sharp, and twisting will break them. When chamber temperatures are too high, you will find gum and varnish deposits on the piston skirt and in the ring grooves. In extreme cases, varnish will bind the rings into the grooves. Broken rings may be the result of either simple failure or grooves which have pounded themselves out.

Fig. 8-32 Piston pin press. (Courtesy Kohler of Kohler)

Excessive up-and-down clearance allows the rings to twist. Damage by detonation will shatter the carbon on part of the piston crown and, if prolonged, will cause metallurgical changes in the casting. Typically, the affected area on the crown will take on a dull, pounded look and may show hairline fractures. Ultimately, the piston will develop holes. Preignition burns piston crowns rather than hammering them. Vertical scratches on the flanks are caused by oil—contaminated either from the by-products of combustion, or because of bearing failure. In 2-cycles, such scratches point to an inefficient air filter or to rust in the fuel.

The piston should be miked across the lands (Fig. 8-33), and at the skirt. The latter measurement is taken at 90° to the wristpin bore. Many small-engine pistons are tapered to allow for heat expansion above the wristpin. Taper varies with the manufacturer, and with the intended thermal loads, but it is on the order of 0.00125 in. per inch of piston length. Ring groove width is checked with a new ring inserted in the groove as shown in Fig. 8-34. The upper ring groove will receive the

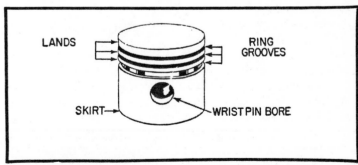

Fig. 8-33 Piston nomenclature (Courtesy Clinton Engines Corp.)

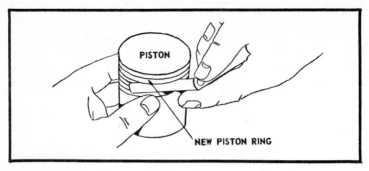

Fig. 8-34 Checking groove clearance—use a new ring (Courtesy Briggs & Stratton)

worst pounding. In general, the piston should be rejected if the gap exceeds 0.005 in. on 4-cycles, and 0.004 in. on 2-cycles. This specification varies somewhat between engine makers. Briggs & Stratton, for example, allows 0.007 in. of gap.

The ring grooves may be cleaned with a tool designed for this purpose or with a broken ring inserted in a file handle. Most small engine rings have a very fine crystal structure and break squarely. Two-cycle rings may be ground to obtain a cutting edge.

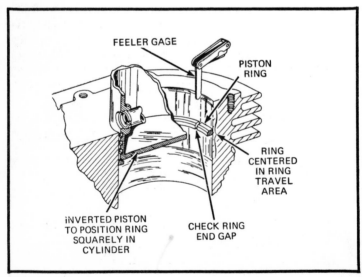

Fig. 8-35 Checking ring end gap. Use the piston to guide the ring parallel to the bore and take a measurement at some point near the center of ring travel (Courtesy Tecumseh Products Co.)

Rings

Reusing piston rings is poor economy. So much time and labor is involved in teardown that the 2 or 3 dollars saved on a new ring set becomes inconsequential, even if the old rings show little wear. (An index of the latter is the gap between the ring ends when the ring is inserted into the bore.)

Ring sets are available in the standard 0.010, 0.020, and 0.030 in. oversizes. These rings must be used with correspondingly oversized pistons and bores. Special ring sets are available from the major manufacturers which have more elasticity than the originals, and which have expanders to increase ring tension. These ring sets conform to cylinders which have as much as 0.008 in. taper and out-of-round. Engineered rings (to distinguish them from factory or OEM rings) are generally limited to standard and first overbore.

Measure the end gap before installing the rings on the piston. Too little gap means that the ring ends will butt under heat and destroy the bore; too much means excessive blowby. As a rule of thumb, the gap should be 0.002 in. per inch of cylinder bore diameter. This rule is subject to modification in high-performance engines which may have slightly more gap, and for low-speed clunkers which can get along with less. Insert the individual rings in the bore as shown in Fig. 8-35. The purpose of using the piston as a ram is to insure that the ring is at right angles to the cylinder walls. Measure the gap

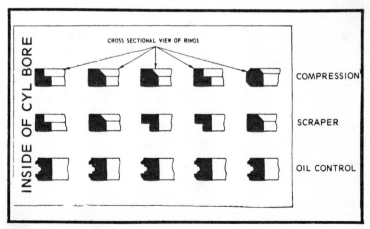

Fig. 8-36 Ring profiles. Other scraper rings may have a bevel on both outer corners (Courtesy Briggs & Stratton)

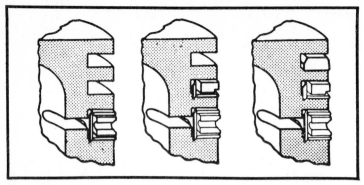

Fig. 8-37 Engineered ring set installation sequence (Courtesy Clinton Engines Corp.)

with a feeler gage. The gap can be enlarged with a file. Mount the file in a vise and move the ring across it, keeping the cut at 90° to the ring face.

Rings have a definite top and bottom—except for some 2-cycle types. Wrong installation will result in loss of power and excessive oil burning. Some rings are marked for position and orientation, but don't count on it. Figure 8-36 shows ring profiles for Briggs engines; these profiles are typical of 4-cycle practice. Figure 8-37 illustrates the assembly sequence for Clinton engineered rings. Begin work from the bottom with all ring sets. Install the oil ring expander first, then spiral the lower rail in the groove. Install the spacer above the rail and spiral the second rail over it. Stagger the ends of the two rails and cast-iron spacer. Next, install the expander in the scraper ring groove. Then spiral the chrome rail into the bottom of the groove. Install the cast-iron scraper ring, with the bevel down as shown in B of the drawing. Finally, insert the compression ring, inner bevel up.

Some 2-cycles have pegs in the ring grooves to prevent the rings from rotating. (A ring end could snag on the ports.) Locate the rings carefully on these pins. Other pistons have floating rings. Assemble so that the gaps are staggered as far apart as possible. A three-ring piston should have the gaps staggered 120 degrees; a two-ring piston, 180 degrees.

Connecting Rods

Figure 8-38 (A through G) is a portfolio of connecting rod disasters. The rod at A overheated and welded itself to the crankpin journal a few thousandths of a second before it snapped. The rod at B was in the process of disintegration; it shows the telltale blue marks associated with high heat. These two rods are both tin-plated. Cast aluminum and bronze rods will not show this color change; you may, however, find score and smear marks on the journal. The proximate cause of overheating is lack of oil, improper oil, or oil which has lost its lubricating qualities through contamination. The ultimate cause may be traced to ignition timing, lean air/fuel ratios, clogged cooling fins, or failure to change the oil at the recommended 25 hr intervals.

Undertightening produces the results at C. Note the radial dispersion of cracks around the cap. Primary shaking forces, unrestrained by rod—bolt tension, literally exploded the rod. The damage at D is believed to have been caused by the lockwashers taking a set. The washers must have spring to bite into the cap metal. Otherwise, the bolts can vibrate loose. The effect of overtightening is shown at E. The aluminum on the bolt threads indicates that they were stripped. The next drawing shows the results of lubrication failure caused by a broken dipper, and G illustrates what happens when the oil is contaminated.

Breakage due to excessive engine speed is not shown in these pictures; this condition is caused by a runaway governor and, assuming the lubrication has kept up with loads, takes the form of a simple fracture near the midpoint of the shank.

The rod journal can be checked with a micrometer, a hole gage, or a plastic gage. The latter method, one of the most often used by small-engine mechanics, involves the use of a plastic wire with a very precisely controlled diameter. The gage wire is placed on a dry crankpin as shown in the next two illustrations. Make the initial measurements on the area of the bearing which is below and in line with the piston at TDC. This is the area which suffers the most wear. Dry the parts thoroughly and torque the capscrews to specified values in small increments. Do not allow the crankshaft to rotate. Remove the cap and compare the width of the plastic gage against divisions on the paper container. This method is fully as accurate as a micrometer and much less expensive. Laying the gage wire longitudinally as in Fig. 8-39 gives the running clearance; gaging across the journal as in Fig. 8-40 gives taper. Out-of-round can be determined by rotating the crank 90° to the bore.

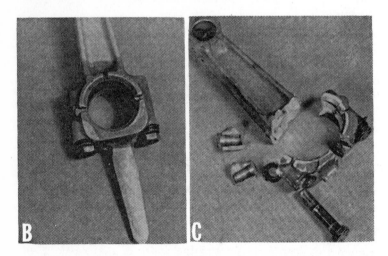

Fig. 8-38 Rod failures A. Rod failure due to heat and allied loss of lubrication B. Overheating (tin-plated rods only) C. Undertightening

Rod bearing clearances are most critical. The bearing should have at least 0.001 in. for lubrication, but more than 0.003 in. is reason for rejection. Some of the larger engines can tolerate 0.0035 in. without excessive wear. Out-of-round should

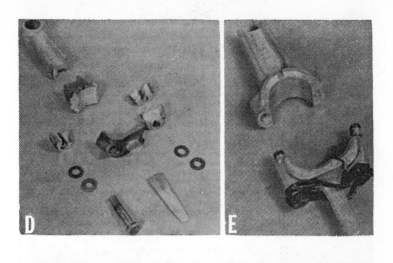

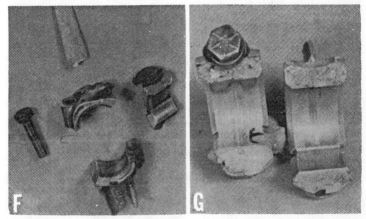

Fig. 8-38 D. Loss of torque because of failed lockwashers E. Overtightening D. Broken dipper G. Object imbedded in cap.

be held to 0.0005 in., and taper to 0.001 in. or less. Do not file-fit a small engine rod.

Thus far we have limited our discussion to plain bearing rods, which account for more than 90% of lawnmower engine production. There are a few well engineered 2-cycles with single- or double-row needle bearings at the big end. The rods are steel forgings and should be inspected for chatter marks. Small imperfections can be remedied by honing. A plastic-gage cannot, of course, be used with antifriction bearings. The rod and crankpin must be miked, although

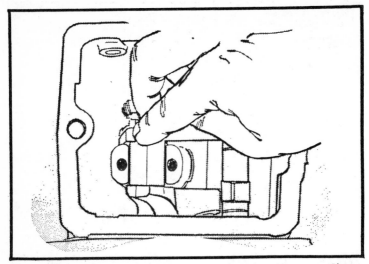

Fig. 8-39 Plastic gage placed for running clearance (Courtesy Clinton Engines Corp.)

experienced mechanics can detect "slap" visually and by touch.

Piston and Rod Assembly

Oil the wristpin bores and insert the pin—preferably with a pin press. Some old-line mechanics, from a more leisurely

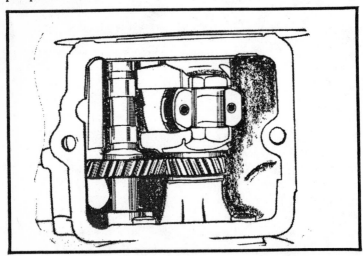

Fig. 8-40 Plastic gage placed for taper (Courtesy Clinton Engines Corp.)

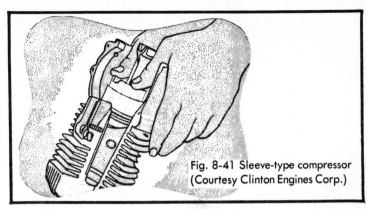

Fig. 8-41 Sleeve-type compressor
(Courtesy Clinton Engines Corp.)

tradition, heat the piston. Flat-topped pistons can be heated, crown down, on a hotplate. A less elegant method is to wrap the piston in a rag soaked with hot oil. If you are very careful, you can play a propane torch over the piston pin bores. Keep the torch moving in small arcs and heat only enough to fit the pin.

Oil the piston liberally; the surest method is to dip the piston wristpin in a bucket of oil. Check the reference marks on the piston and rod and turn the crankshaft to bottom dead center. If the rod has fixed bolts or studs extending downward from the shank (a common design), cover the ends with tape or short lengths of fuel line. Contact with the crankpin journal can create a pit which will have to be smoothed, as best as possible, with emery cloth.

The rings must be compressed to fit the bore. You should have a small-engine ring compressor for this job, although rings have been installed with the aid of three screwdrivers and much exertion. These types of compressors are available; the open or removable type was illustrated in Chapter 3. It is used primarily on ¼ + cycle engines. The cylinder jug must be fitted over the piston trapping the compressor between the crankcase and the cylinder. A sleeve compressor (not absolutely necessary) for Clinton 2-cycle engines is shown in Fig. 8-41. The compressor shown in Fig. 8-42, available from Briggs & Stratton, it has projections on one end for use on cast iron engines which have a pronounced chamfer at the top of the bore. Tighten the compressor to the limit of leverage supplied by the wrench. It helps to hold the assembly by the rod, in a vise. Drive the piston into the bore with a hammer handle (Fig. 8-43).

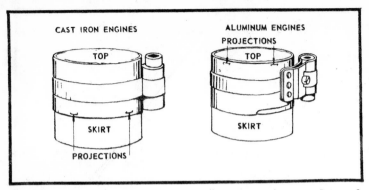

CAST IRON ENGINES

ALUMINUM ENGINES

Fig. 8-42 Standard compressor for small engines (Courtesy Briggs & Stratton)

Drive firmly, but with discretion. If the piston stops, do not force it—a ring has escaped the compressor at the top of the bore. Remove the piston (using a wooden punch) from the bottom and repeat the process. Take plenty of time, since a mistake will result in a broken ring which will not be discovered until the engine is started. (Two-cycle rings can be

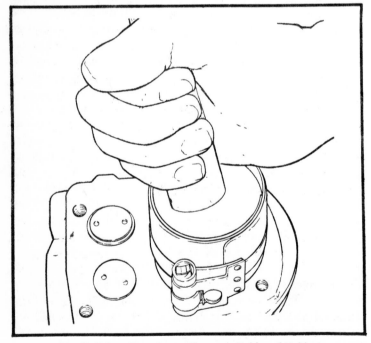

Fig. 8-43 Installing piston (Courtesy Kohler of Kohler)

Fig. 8-44 Grease needle bearings to hold them in place (Courtesy OMC)

checked at the exhaust port with a screwdriver blade—the rings should spring back as they are compressed into their grooves.)

Make certain the upper rod journal mates to the crankpin. The last ¼ in. or so of travel is best taken up with rod bolts. Remembering the horror pictures in Fig. 8-38 should be encouragement enough to use new lockwashers, locknuts, or bolt lock straps. Briggs employs straps, while Lauson and OMC use locknuts, and Kohler lockwashers.

Needle bearings are assembled with the aid of grease or beeswax as shown in Fig. 8-44. Power Products double-row bearings have their flat ends butted (Fig. 8-45). Do not mix old and new bearings; the new bearings are microscopically larger than the old ones and, consequently, will support the full load.

Tin-coated rods become oxidized on the shelf. Remove this oxidation from the bearing surfaces with a clean rag. Oil the journal liberally and fit the cap, noting the match markers. Torque to specification (see Tables 8-1 and 8-2) in several increments. Observe the sequences shown in Figs. 8-46 to 8-51.

New rings will drag, but the crankshaft should turn without harsh binding or sticking. If the cam is in place, you will be able to feel the resistance of the valve springs as the followers climb up on the lobes.

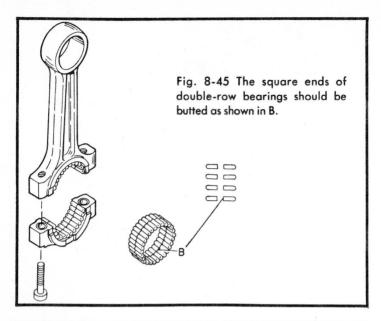

Fig. 8-45 The square ends of double-row bearings should be butted as shown in B.

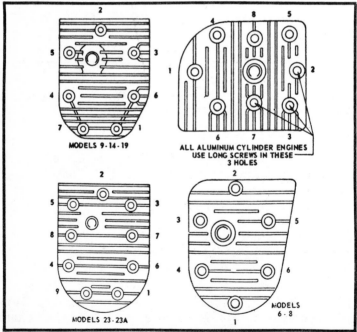

MODELS 9-14-19

ALL ALUMINUM CYLINDER ENGINES
USE LONG SCREWS IN THESE
3 HOLES

MODELS 23-23A

MODELS 6-8

Fig. 8-46 Torque sequence for Briggs heads (Courtesy Briggs & Stratton)

302

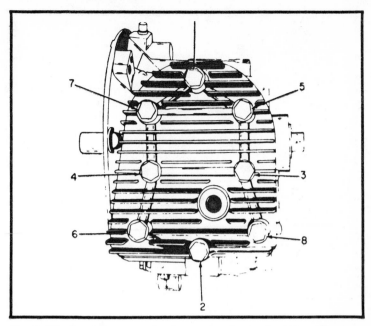

Fig. 8-47A Torque sequence for Lauson heads—under 8 hp (Courtesy Tecumseh Products Co.)

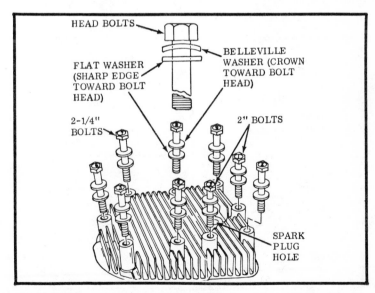

Fig. 8-47B Torque sequence for Lauson heads—8 hp and over: 1, 6, 4, 3, 2, 5, 8, 9, 7.

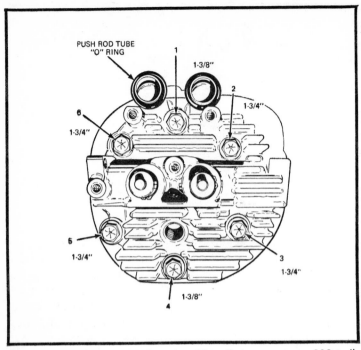

Fig. 8-48 Tighten—1, 3, 5, 2, 4, 6 in 50 in.-lb increments to 200 in-lb
Lauson HH 150 (Courtesy Tecumseh Products Co.)

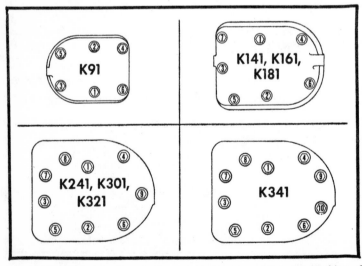

Fig. 8-49 Kohler single-cylinder torque sequence (Courtesy Kohler of
Kohler)

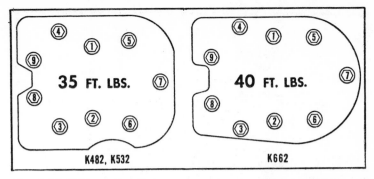

Fig. 8-50 Kohler twin-cylinder torque sequence (Courtesy Kohler of Kohler)

Camshafts

The cam has multiple functions. In addition to opening the valves, it may drive the magneto (in which case it usually incorporates an ignition advance mechanism), unseat the exhaust valve during cranking to reduce compression, drive the oil pump or slinger, and be used as a power takeoff for the propulsion mechanism. Engines intended for long, faithful service employ an axle to support the cam or, more rarely, run the cam on bronze bushings pressed into the block.

It is always wise to find the timing marks before disassembly. These marks index the cam and valves relative to the crankshaft. Early engines did not always have marks; or if they were present, camshaft or crankshaft wear might have obliterated them. Timing marks are usually indentations found on both parts. Traditionally, a pair of indentations on the cam gear lie aside similar marks on the crankshaft gear (Fig.

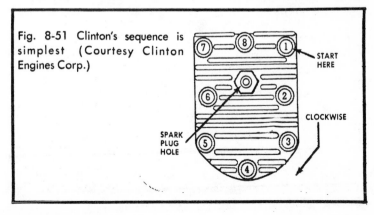

Fig. 8-51 Clinton's sequence is simplest (Courtesy Clinton Engines Corp.)

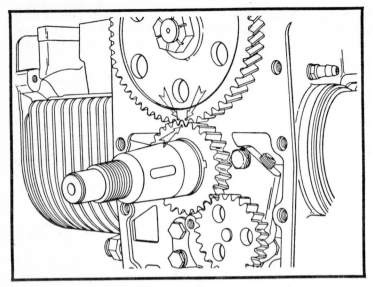

Fig. 8-52 K482 and K532 timing marks (Courtesy Kohler of Kohler)

8-52). Other engines have a single mark at the root of both teeth. Lauson engines use a conventional timing mark on pressed-on crankshaft gears. Gears which are fixed with a key employ the keyway as a reference mark. The drawing in Fig. 8-53 illustrates an exception to the general rule that timing

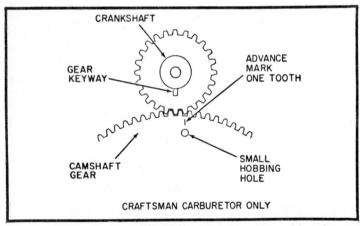

Fig. 8-53 Craftsman Tecumseh-built engine marks (suction-lift carburetor only) do not align. This is an exception and points out the need to discover the timing mark position before teardown (Courtesy Tecumseh Products Co.)

marks align. Apparently because of the characteristics of this carburetor, Tecumseh specifies that the cam be advanced one notch. Another variant is the Briggs design shown in the next figure. Because the crankshaft ball bearing is in the way, the engine is timed using a mark on the shaft counterweight. Other designs with antifriction mains employ a chamfered tooth.

Because of the impracticality of disassembly before the timing marks are located, you will have to scribe your own if you can't find any. Timing "cold turkey" is very difficult and time-consuming since manufacturers do not supply valve lift data.

Inspect the both gears for wear and tooth damage. The lobes normally outlast every other friction surface on the engine, but they are only surface-hardened. Cracks on the surface quickly expose the soft metal underneath, and the lobe rounds off. The advance mechanism may be timed to the shaft and have its own reference mark (Fig. 8-54). Check the weights for free movement.

Camshaft end play is ignored on most engines. The practice of using flange gaskets of various thicknesses to control crank float insures, to some degree at least, that the cam has proper clearance. Kohler provides spacers (Fig. 8-55) at the magneto end of the cam, an unusual practice.

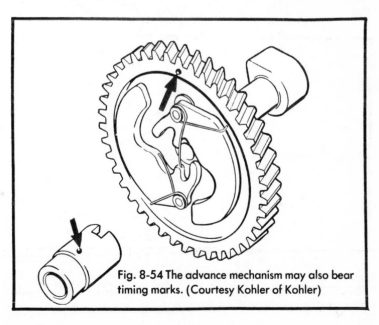

Fig. 8-54 The advance mechanism may also bear timing marks. (Courtesy Kohler of Kohler)

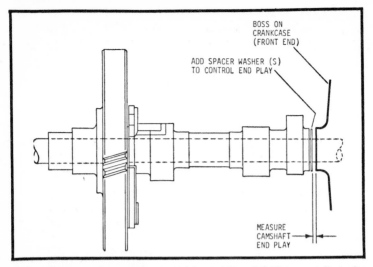

Fig. 8-55 Camshaft end play should be 0.005 – 0.010 in. on all single cylinder engines except the K91, which is 0.005 – 0.020 in. Twin-cylinder camshaft play is 0.017 – 0.038 in. Adjust using spacers of various thicknesses. (Courtesy Kohler of Kohler)

Lubrication Systems

The bulk of Briggs products employ a slinger (Fig. 8-56) to distribute oil around the crankcase. Some models do not have the spring washer in the illustration. Inspect the bearing and

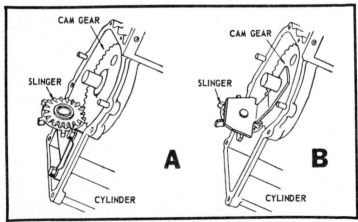

Fig. 8-56 A. Old-style slinger—replace bracket if worn to a diameter of 0.49 in. B. Late slinger. Some models have a wave washer over the cam gear stub between the slinger and the flange (Courtesy Briggs & Stratton)

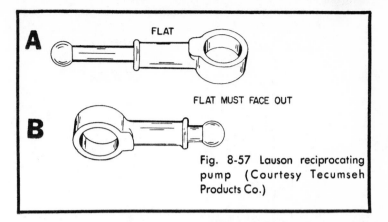

Fig. 8-57 Lauson reciprocating pump (Courtesy Tecumseh Products Co.)

the teeth for wear. Other engines employ scoops on the connecting-rod cap, or on the camshaft. Lauson is famous for their reciprocating oil pump; it is driven by an eccentric gear on the camshaft. The flat side is out, away from the engine (Fig. 8-57). Oil is drawn into the hollow camshaft on the intake cycle of the cam. As the pump telescopes, a second port in the camshaft hub is aligned with the barrel, and oil passes through it to the top end of the camshaft. From there it goes to the upper main bearing and through the crankshaft to the rod bearing. A pressure-relief valve is incorporated in the gallery which opens when the oil is thick (cold) or when the system clogs. Normal pressure is 7 psi. Figure 8-58 shows the oil circuit. In addition to oiling by the pump, the engine receives lubrication by splash. Some engines had an air-bleed port in the gallery to atomize the oil going to the upper main bearing. These ports should be blown out, and the pumps primed, prior to assembly. Do not change the adjustment on the relief valve.

Other Lauson engines employ a gear-driven pump as shown in the next illustration. Inspect the pump for wear and scuffing, and replace as an assembly if you have the slightest suspicion that all is not right. The clearance between the cover and oil pump gear can be determined with a feeler gage. It should be between 0.006 and 0.007 in.; the distribution system is similar to the one employed with the reciprocating pump. Flood the parts with oil immediately prior to assembly to insure early oil delivery.

The larger engines in this family employ a slinger on the crankshaft. In the event the main bearing is removed, it is vital to make sure that the slinger is not installed 180°

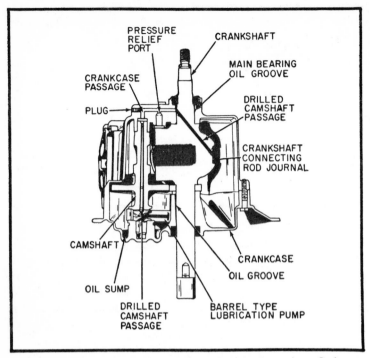

Fig. 8-58 Lauson oil circuit (Courtesy Tecumseh Products Co.)

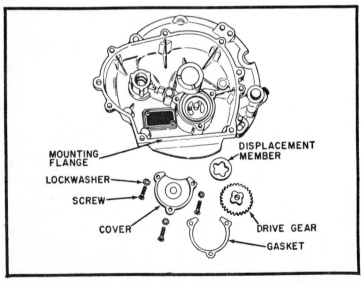

Fig. 8-59 Lauson geared pump (Courtesy Tecumseh Products Co.)

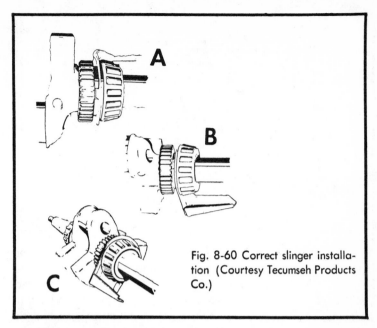

Fig. 8-60 Correct slinger installation (Courtesy Tecumseh Products Co.)

out-of-phase. The three views in Fig. 8-60 show the slinger correctly installed with the scoop on the counterweight side.

Figure 8-61 illustrates the Clinton oiling system. The geared pump delivers oil through the camshaft to an exhaust port near the upper main bearing. The lower parts of the engine are lubricated by splash.

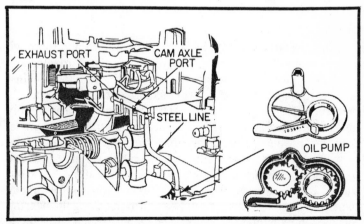

Fig. 8-61 Clinton pump and lubrication circuit (Courtesy Clinton Engines Corp.)

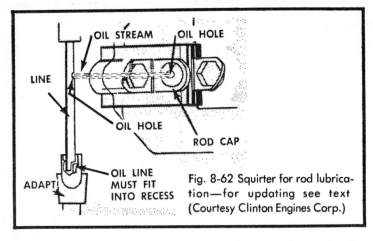

Fig. 8-62 Squirter for rod lubrication—for updating see text
(Courtesy Clinton Engines Corp.)

Labels in figure: OIL STREAM, OIL HOLE, LINE, OIL HOLE, ROD CAP, ADAPT..., OIL LINE MUST FIT INTO RECESS

An interesting variation was used on early production pumpers—Clinton apparently had some scoop rods in stock and modified the delivery line to squirt oil into the scooper. Replacement rods have been designed to attract adequate lubrication from the main bearing, and the spray nozzle is no longer needed. If you have one of these engines and have changed the rod, it is suggested that you replace the oil line with part No. 158-52. This line delivers all of the output to the main bearing. You will need pump adapter No. 1-10 for this changeover. However, if you replace the pump on one of the squirter engines, you must also change the line and adapter to the new specifications.

Crankshafts

The crankshaft should be given a complete and thorough examination. To do it right you need a pair of machinist's **V**-blocks and a micrometer (for the main bearings). Most shops are satisfied if the power-takeoff end is not obviously bent (see Chapter 3) and the bearing journals are reasonably smooth. Crankshafts can be straightened with an arbor press, but *the practice should be discouraged* because of the possibility of surface and subsurface cracks which could lead to serious mechanical damage **and personal injury**. Blow out the oil passages and polish the journals with No. 600 emery paper or stripping. There should be no ridges along the bearing journals. Small imperfections can be smoothed out with emery cloth, but be careful not to undersize or flatten the shaft (Fig. 8-63).

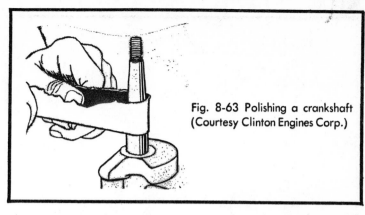

Fig. 8-63 Polishing a crankshaft
(Courtesy Clinton Engines Corp.)

Crankshaft end play can be determined with a feeler gage or, more accurately, with a dial indicator. The shaft is installed in the block and bumped with a mallet, or levered, to the extremes of fore and aft travel. Figure 8-64 illustrates the procedure on a Kohler twin. In this case the upper bearing

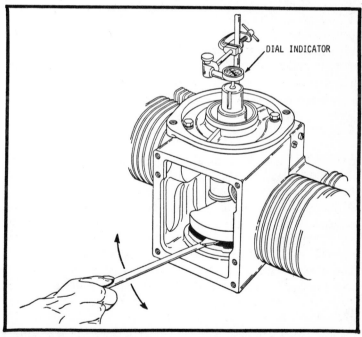

DIAL INDICATOR

Fig. 8-64 Checking crankshaft end play with a dial gage. If you use a feeler, measure between the counterweight cheeks or the pulley or the power-takeoff pulley and block (Courtesy Kohler of Kohler)

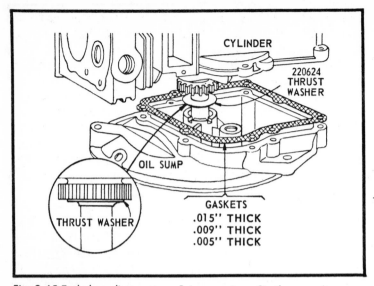

Fig. 8-65 End play adjustments on Briggs engines. Similar expedients are employed by other manufacturers (Courtesy Briggs & Stratton)

plate gasket is critical. The manufacturer supplies gaskets and shim sets for the engine in varying thicknesses. The end play should be 0.004 to 0.010 in. maximum. Briggs & Stratton employs a similar system on their products. Bearing plates and several flange gaskets from 0.015 in. (standard) to 0.009 in. are available, as well as shims. End play should be between 0.002 and 0.008 in. on all models. Use the standard gasket on aluminum blocks to prevent possible oil leakage. If a new flange has been installed, end play may be less than desirable. Wear on the crank and thrust bearing surfaces will cause endplay to be greater than specified. The thrust washer in Fig. 8-65 is placed at the power-takeoff end on aluminum engines with bushings; aluminum blocks with ball bearings take a shim on the magneto end.

The same procedure is used on other 4-cycle engines. Where there is provision for end play adjustment on 2-cycles, shims are employed at the magneto end of the crankshaft.

Main Bearings

Most power mower engines support their crankshafts on plain bearings. Iron blocks are fitted with bushings at either end. Aluminum blocks may have a bushing at the magneto end but, more often than not, are entirely without bearings as such.

The block material is compatible with the cast-iron crankshaft. Heavy-duty engines are distinguished by antifriction main bearings.

Plain-bearing engines, whether originally fitted with a bushing or not, can be rebushed with factory-supplied reamers and guide blocks. It would be impractical for the individual owner to purchase this tooling since the cost exceeds the price of an easily available short block; and small repair shops would find the reamers rusting between calls, because few customers realize the extent of small engine repairability. But engine dealers and distributors have these tools as part of their working inventory.

The sequence of illustrations beginning with Fig. 8-66 details the process for a light-frame block which had aluminum bearing surfaces. Assemble the flange and, using the large-diameter cutter, ream the bearing oversize to accept the bushing. Use kerosene as a lubricant, and turn the reamer clockwise. Counterclockwise turning will dull this or any other cutting tool. Remove the flange and install the bushing with

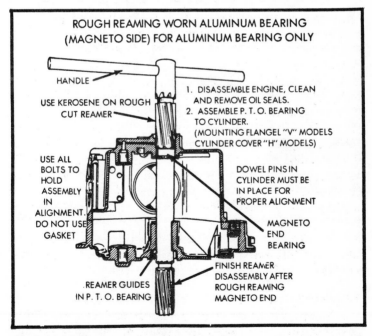

Fig. 8-66 Rough reaming the magneto side bearing (Courtesy Tecumseh Products Co.)

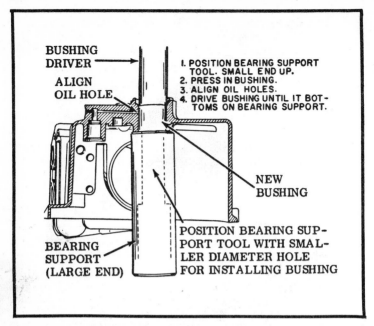

BUSHING DRIVER

ALIGN OIL HOLE

NEW BUSHING

BEARING SUPPORT (LARGE END)

POSITION BEARING SUPPORT TOOL WITH SMALLER DIAMETER HOLE FOR INSTALLING BUSHING

Fig. 8-67 Installing magneto side bushing. The bearing support tool established the depth of press. Other manufacturers give this dimension as a specification (Courtesy Tecumseh Products Co.)

the aid of a bearing support (Fig. 8-67). Installation procedures vary somewhat between manufacturers. The critical points are *depth of the bushing* and *alignment with the oil supply port.* If installation of a power-takeoff (PTO) bushing is desired, assemble the flange (or bearing plate) and rough-cut as shown, using the new magneto bushing as a guide (Fig. 8-68). Install the PTO bushing to the prescribed depth and finish-ream.

The job is simpler on engines which have factory-installed bushings; the old bearings are driven out and new ones are pressed into place. Then both are finish-reamed. Ideally, this process should be used whenever a new flange or bearing plate is installed, to insure crankshaft alignment.

Antifriction bearings are, as they say, another ballgame. Typically, the bearing consists of balls or rollers held between two races. The inner race is an interference-fit to the crankshaft, and the other race may be pinned into place (Briggs & Stratton, Tecumseh, Wisconsin, some Clinton) or interference-fitted. The latter method is widely employed in

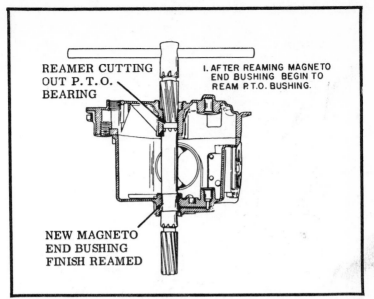

REAMER CUTTING
OUT P.T.O.
BEARING

I. AFTER REAMING MAGNETO
END BUSHING BEGIN TO
REAM P.T.O. BUSHING.

NEW MAGNETO
END BUSHING
FINISH REAMED

Fig. 8-68 Rough remaining power-takeoff bearing (Courtesy Tecumseh Products Co.)

2-cycle engines and in those from Kohler. In general, the outer race is somewhat looser than the inner race, and cases or bearing plates can be parted with a mallet struck against the crankshaft. The bearing plates and crankcase halves for Kohler engines should be disassembled with a puller, and may require heat as well (Fig. 8-69).

Do not remove the bearings from the crankshaft unless they show evidence of failure. Look for signs of breaks in the polished surface of the races. If the races are concealed by the bearing retainer, remove all lubricant and rotate the bearing by hand. Do not use a compressed-air gun. Noise or roughness means that the bearing should be replaced.

Score marks on the outside diameter of the outer race are irrefutable evidence that the bearing has "turned." If the condition has not persisted over a long period, the clearance between the bearing plate (or crankcase half) may be close to specifications; the bearing can be secured with a Loc-Tite bearing mount. If this isn't possible, the parts will have to be replaced.

Remove the bearing with a splitter (Fig. 8-70). Install the new bearings with a suitable driver or arbor press. Heat the bearing in oil to reduce assembly stresses. The bearing should

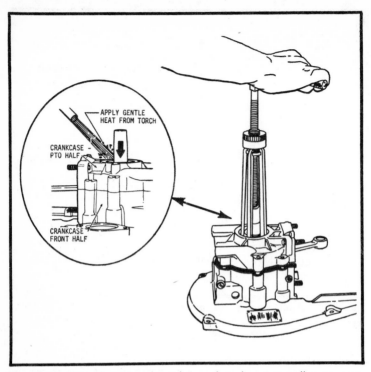

Fig. 8-69 Using heat (moderately) and a bearing puller to part crankcase halves on a 2-cycle engine (Courtesy Kohler of Kohler)

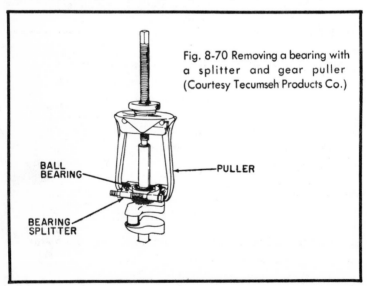

Fig. 8-70 Removing a bearing with a splitter and gear puller (Courtesy Tecumseh Products Co.)

Fig. 8-71 Supporting crankpin during bearing installation (Courtesy Rockwell Mfg. Co.)

be placed in a container half-filled with motor oil and supported by a wire to prevent contact with the container's bottom or sides. Temperature should not exceed 325°F. It is wise to block the crankshaft cheeks, as shown in the next drawing, to prevent crankpin buckling. Kohler bearings are first pressed into the bearing plate, and then the plate is pressed over the crankshaft. (Fig. 8-72 A and B.)

You can purchase antifriction bearings from supply houses which specialize in this trade; in some cases you may be able to save money by doing so. But, a word of caution: Be sure that the replacement matches the original in all significant respects; do not go to the standard C1 clearance for bearings with pressed inner races. The race grows enough to fill this clearance and is subject to premature failure from lack of lubrication. Most small engine main bearings come under C3 or C4 specifications.

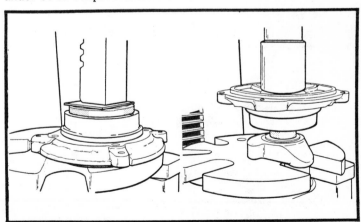

Fig. 8-72 A. Pressing bearing into closure plate. B. Pressing plate on (supported) crankshaft (Courtesy Kohler of Kohler)

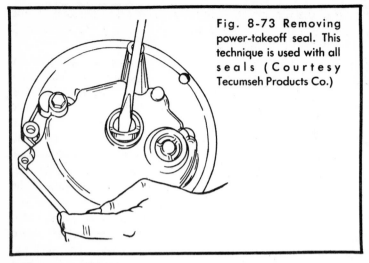

Fig. 8-73 Removing power-takeoff seal. This technique is used with all seals (Courtesy Tecumseh Products Co.)

Seals

Oil seals are removed per the example in Fig. 8-73. With practice you can pop out seals as easily as opening a can of beer. Many Lauson engines have a washer-like dust cover over the seal at the PTO end. This cover must be pried off before the seal can be removed.

Look the seals over for leakage, cuts on the lip, deformation, and for loss of elasticity. It is suggested that you polish out wear marks on the crankshaft which would accelerate seal failure. Careful mechanics install the shaft with the aid of a seal loader (as shown in Fig. 8-74). Seal loaders for various crankshaft diameters are available from

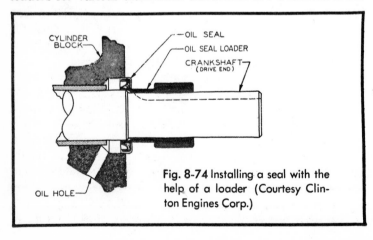

Fig. 8-74 Installing a seal with the help of a loader (Courtesy Clinton Engines Corp.)

engine distributors or directly from the factory. Some protection, though, can be had by covering the crankshaft keyways with masking tape.

Seal installation is not difficult—the steep edge of the lip goes toward the pressure side. Oil or crankcase pressure acts on this steep edge to force the seal tighter against the crankshaft in a manner analogous to that of chamber pressure pushing against piston rings. The outer diameter of the seal may be coated with Neoprene to make it conform to small distortions in the seal boss. If the replacement part has a steel back, coat it—not the lip—with gasket cement. Drive the seal in place with a suitable tool. Again, these tools are made at the factory and should be used whenever available. In a pinch, you can substitute a length of pipe (cut off square), whose outer diameter bears against the flanged part of the seal. Pressure on the middle of the seal will distort it. Drive it to the original depth, observing any oil drain holes which must be clear, and inside of the sealed area.

Cylinder Bores

Power mower blocks have either cast-iron or chrome bores. The latter was pioneered by Briggs & Stratton. A thin layer of chrome is applied directly to the aluminum block metal and reverse-etched to form a semiporous finish for oil retention. Since no ferrous liner is required, these engines run cooler than their more conventional brethren. Bore wear is almost nil because of the great difference in hardness between chrome and cast-iron rings, and because of chrome's resistance to acid. On the other hand, these engines tend to complicate the supply picture since plated pistons are required. Reboring is out of the question. Inspect the finish for evidence of deep scratches and peeling. If the chrome is in the progressive condition of peeling, replace the block.

Cast-iron bores should be honed whenever new rings are installed. Honing amounts to deliberately roughening the cylinder to encourage wear. Cylinder walls take on a glaze in service; the surface metal compresses and smears into a glass-hard surface. New rings conform to the bore through a mutual wearing process; this requires that the bore be rough. Eventually, rings will seat to a glaze, but the engine might wear out before it stops burning oil.

Glaze busting requires more than a ¼ in. drill motor and a brake cylinder hone—although this is the extent of

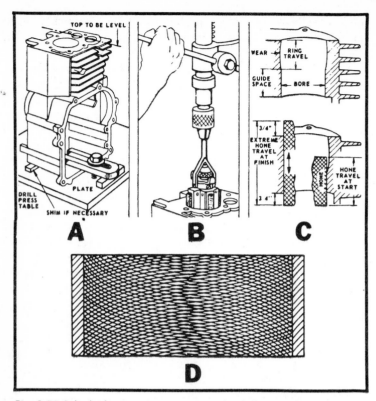

Fig. 8-75 Cylinder honing. A. Mounting on the drill press table—cylinder must be vertical. B. Upper limit of stroke; hone protrudes a like amount at the bottom limit. C. Begin oversizing operations at the point of least bore wear to guide the hone. D. The desired crosshatch pattern for both deglazing and honing oversize (Courtesy Briggs & Stratton)

sophistication for some shops. One needs a hone designed for this purpose, the appropriate stones, and a heavy drill motor or press. The finished surface should have the crosshatch pattern shown in Fig. 8-75D. To arrive at this, we need a spindle speed of 400 rpm or less, and a lateral speed of about 70 strokes per minute. The low spindle speed effectively rules out small drill motors unless, of course, the drill has a speed control.

Stones are available according to two standards—grit numbers and series number. The former refers to how many particles occupy 1 square inch. Series numbers have superseded the older grit-number standards, and are used by most manufacturers today. (A Series 500 hone has a No. 280

grit.) As a rule, the harder the ring material, the finer the hone required. Cast-iron rings seat best with a series 200 finish, chrome with a 300, and stainless with a 500 finish. Some leeway is allowed in the choice of whether or not to lubricate. Dry-honed cylinders will be slightly rougher than those which were oiled. However, once a hone has been wetted, it must continue to be used lubricated. Use animal fat or vegetable oil—petroleum based oils dissolve the binder holding the stone particles together.

Keep the hone moving at a constant speed, and extended ¾ in. beyond the hone limits. Do not pause at the end of the stroke; keep in mind the fact that a hone cuts with each revolution. As soon as the surface glaze is removed, you can stop. Stone particles are more difficult to remove from the pores of metal than is generally realized; solvents merely float them in deeper where they continue to abrade the rings and piston. The best method is to scrub the cylinder with hot water and detergent, followed by a drying with paper towels. As long as the towels discolor, particles are still present.Oil the bore as soon as you are satisified that it is clean.

Cylinder bores suffer from out-of-roundness, wear-induced oversize and taper, and deep, localized scratches. A quick check of taper can be made by inserting two new piston rings in the bore (Fig. 8-76). Use the piston as a guide to square the

Fig. 8-76 Using rings to determine cylinder wear (Courtesy Clinton Engines Corp.)

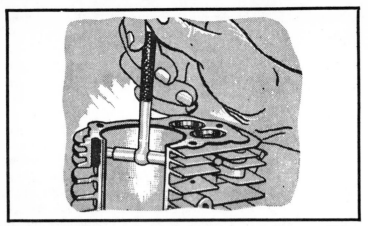

Fig. 8-77 A more accurate method of sizing the bore (Courtesy Clinton Corp.)

rings. The difference in the end gaps reflects the amount of taper. A more accurate method is to use a hole gage at the extremes of ring travel. Turn the gage 90° to detect out-of-round. Specifications vary with bore diameter and use-profile, but 0.003 in. oversize, 0.002 in. taper, and 0.005 in. out-of-round are good conservative limits.

Cast iron cylinders can be honed oversize to accept larger piston and ring sets. These parts are available in 0.010 in., 0.020 in., 0.030 in., and 0.040 in. increments over standard. Piston crowns are marked with the numerical oversize.

Place the block under a drill press as shown in Fig. 8-75. Do not bolt the block down; some free movement will aid centering. Use coarse roughing stones for initial honing. Determine the oversize by miking the replacement piston. Some manufacturers provide a running clearance for oversize pistons—others do not. Begin at the bottom of the bore (where it should be truest) and progress on each stroke toward the top. Measure the cylinder frequently to determine the speed of cutting. Continue in this manner until you are within 0.002 in. of the desired size, then change to a finishing stone. Clean the bore as described previously.

It is also possible—some would say preferable—to bore a cylinder on a lathe or boring machine. This method insures accuracy if the block is mounted rigidly to the tool table and the cutter is sharp. A dull cutter "threads" the cylinder, leaving an entirely inadequate surface.

Chapter 9

Drive Systems

Internal combustion engines develop torque at relatively high rotational speeds and stall when subjected to heavy loads, making some sort of speed—torque conversion necessary. Most power transmission devices reduce engine output speed while at the same time increasing available torque. The more sophisticated mowers incorporate differential gears on the driving wheels to prevent scalping as the inner wheel attempts to match the speed of the outboard wheel during a turn. Power transmission devices can be grouped according to their principles of operation into shaft-, belt-, chain-, or gear-driven types.

Shaft Drives

The most frequently used form of power transmission is shaft drive. In essence, a drive shaft provides an *engine speed* to *torque* ratio of only 1:1. However, in the case of rotary lawnmowers, there is power transfer from the point of view blade tip velocity. The blade tip, at the end of a lever, dissipates torque in the form of velocity (which can be as high as 200 mph at full engine speed). The smaller the blade, the lower the tip velocity—hence, better conservation of torque. If other factors are equal, a machine with a small-diameter blade will accept heavier cutting loads than one with a larger diameter blade.

In the classic rotary mower, the blade is bolted directly to the crankshaft stub through an intermediate adapter or flange. Some adapters have shear pins cast-in, which mate with holes in the blade. In the event of a sudden overload, the shear pins will protect the blade and crankshaft. A less destructive method is illustrated in Fig. 9-1. Washers (19), made of Neoprene, are preloaded by the bolt (3) acting against the steel washer (2). The tension on the bolt should be periodically

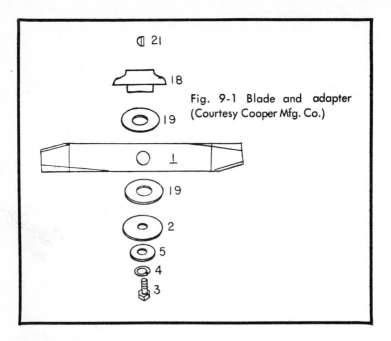

Fig. 9-1 Blade and adapter (Courtesy Cooper Mfg. Co.)

checked, and adjusted if necessary, to compensate for compression of the washers.

Another variant is shown in Fig. 9-2. Controlled slippage is provided by the dogpoint screws between the inner and outer adapter housings. Since the inner housing is rimmed, loss of the blade is virtually impossible.

Blades, and blade stiffeners, are extremely critical parts and warrant regular inspection. Figure 9-3 illustrates a collection of ills common to all rotary blades; the next figure (9-4), shows those problems typical of which can develop in blade stiffeners.

Servicing these areas should be undertaken with a bit of caution; many mechanics who were unaware of the dangers involved carry scars from rotary-blade mowers. Before removing the blade or beginning any underside repairs,

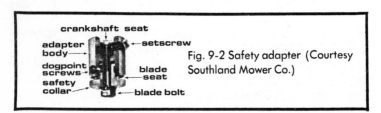

Fig. 9-2 Safety adapter (Courtesy Southland Mower Co.)

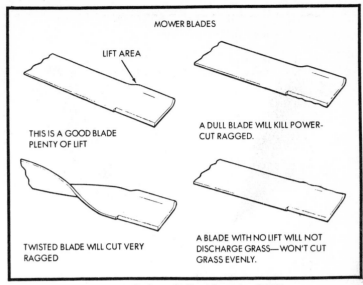

Fig. 9-3 Blade maladies (Courtesy OMC)

disconnect and ground the spark plug lead to the cylinder head. Do not leave the wire dangling—it may come into contact with the spark terminal. Shielded boots can be grounded by inserting a bolt into the lead terminal and then wedging it between the radiator fins. Raise the machine on its front wheels and remove the blade bolt with a box-end wrench; secure the blade with a wood block as shown in Fig. 9-5. Replace the bolt if it has become battered or if its threads show

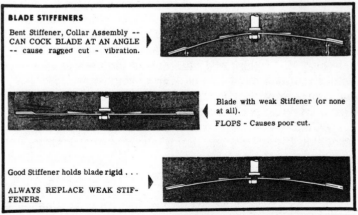

Fig. 9-4 Blade stiffener problems (Courtesy OMC)

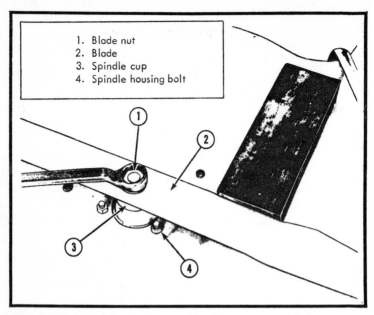

1. Blade nut
2. Blade
3. Spindle cup
4. Spindle housing bolt

Fig. 9-5 Technique for removing and tightening rotary blades using a box-end wrench or socket to minimize the chance of slip (Courtesy International Harvester)

signs of wear. If possible, replace with a factory part from your dealer, or use an SAE grade 8 bolt (see Fig. 8-5B) with the same machine thread, diameter, and length. You can purchase these high-quality bolts from International Harvester and Caterpillar tractor dealers. Replace the lock washer if it has flattened, and inspect the adapter for damage or wear. A loose adapter will make a knocking sound at low speed and may damage the crankshaft.

Sharpen the blade on a grinder, being careful not to overheat it. Blackish and blue marks indicate blade temper loss, a cause of rapid dulling. Unless you are well acquainted with metal working, do not attempt to quench the blade. An overhardened blade is prone to shattering.

Balance the blade with a static balancer; these are available from Sears and many lawnmower parts houses. If possible, purchase an aluminum balancer rather than a plastic one—aluminum wears better and seems to give more consistent results.

A drive shaft should be checked for trueness (Chapter 4), be as short as possible, and have a minimum of overhang on its

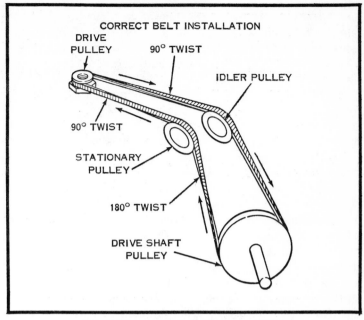

CORRECT BELT INSTALLATION

Fig. 9-6 Serpentine belt (Courtesy OMC)

bearings. This is particularly important in rotary mower drive shafts which are subject to tremendous loads. Other shafts, such as those found in transmissions and reel mower power-takeoff assemblies, can be checked with machinist's V-blocks or by chucking them in a drill press or lathe. The shafts should be true within 0.001 in., but this tolerance is less critical when self-centering bearings are employed.

BELT DRIVES

Pulley belts are quiet-running, relatively inexpensive, and will tolerate misalignment that would quickly destroy gears or chains. Power can be delivered at right angles to the input sheave, or pulley, by means of an idler wheel (Fig. 9-6), and the output can be easily varied by simply changing the sheave ratio. To some small but significant extent, belts compensate for loads by creeping out on the *driven* sheave and snuggling deeper into the *drive* sheave (Fig. 9-7). This feature, plus the inherent shock-absorbing ability of fabric and glass, means prolonged bearing life and reduced stresses on the shafts.

On the other hand, belts are designed to be sacrificed, and require periodic replacement. Inspect the belt carefully, since it will give you clues to machine operation difficulties. Wear on one flank indicates that the sheaves are out of alignment. Local wear spots on both flanks mean that the system has been overloaded. In most cases, a new belt installed with the right tension will correct this problem. A belt that has turned itself over is a sign of extreme sheave misalignment. Do not attempt to use a twisted belt because, most probably, the cords have stretched or parted. Cracks on the inner diameter are caused by heat and usually develop when the belt sides have worn enough to allow slippage. Buffing (dark, shiny marks) on the inner diameter are caused by belt contact with the bottom of the sheave grooves, and usually indicate that the belt, sheaves, or both, have worn beyond safe or efficient use. Wear on the outside circumference can point to a misadjusted belt guard or a frozen idler pulley.

Belts are stock-coded by load rating, width, and length. The width is the dimension across the widest surface and is ⅜ in. for most garden machine applications. Length is the distance measured around the outside edge of both sheaves—not in the groove. These dimensions are encoded on the original belt. Codes vary, but the first digit always refers to width. For Gates belts, 1 means ⅜ in., and 2 indicates ½ in. width. The next three digits describe the belt length in inches. Another popular line goes by eighths of an inch; 3 means ⅜ in., 4 refers to ½ in., and so on. Because of the variation in coding systems, a belt measuring stick is almost a necessity to insure correct replacement size.

Lawnmower and garden equipment belts are specially formulated to resist oil and abrasion damage; most drive belts have steel or fiberglass cords. Ordinary hardware store items—intended for home appliances—are never adequate replacements and will quickly fail in lawnmower service. Get an exact replacement from your dealer or consult a belt manufacturer for a compatible type.

BELT MAINTENANCE

Belts can tolerate some alignment irregularity but are happier when properly adjusted. Center the sheaves using a straightedge placed across the grooves. Check the guards and guides for evidence of rubbing and adjust or bend as necessary. Belt tension may be varied in a number of ways; the most common is to adjust fore and aft movement of one sheave within elongated bolt holes. Another method is to install spacers under one drive element. On a few machines, the throw of the idler sheave arm can be adjusted to make the belt tighter or looser.

These adjustments should be made judiciously, however; too much tension will pit the bearings and cause overheating—too little will result in slippage, belt burns, and squeal. On typical mower and garden machine installations, belt deflection under thumb pressure should be ¼−½ in. either side of taut center. (This figure is approximate and is subject to modification by a manufacturer's specifications and field experience.) Another method, traditionally popular in heavy industrial facilities, is to strike the belt with the palm of the hand; a belt that doesn't yield is too tight. A loose belt has a "dead" feel as it sinks into the grooves. Correct tension produces a "live" or springy response.

When installing a belt, inspect the sheaves; the belt should be level, or no lower than $\frac{1}{16}$ in. below the top of the sheave. Wear on either part will allow the belt to sink into the groove; and in extreme cases, the belt will bottom, depending entirely on tension to transmit torque. Normally, a V-belt operates through a continuous wedging action between its flanks and the groove.

Small sheaves are normally die-cast in zinc pot metal. This material, relatively harder than the belt fabric, does a fair job of resisting wear—however, steel sheaves are the best choice for edgers, tillers, and other applications where abrasion levels are high. The only disadvantage of steel in this application is its tendency to rust.

FRICTION DRIVES

In an absolute sense, a V-belt is a species of friction drive. However, this term is usually reserved for contrarotating components whose point of contact is a frictional surface. Devices based on this drive were popular in the cycle cars of the 1920s and have been used in a variety of applications from gun-laying mechanisms to dental drills.

Figure 9-8 illustrates a friction drive used in the OMC Loafer. The rubber-tired wheel is perpendicular to, and is held in contact with, the revolving disc located below the engine. Speed reduction is achieved by moving the wheel toward the center of the disc where progressively slower moving points of contact are encountered—an infinite number of ratios is available in forward and reverse. These devices require little maintenance because of their innate simplicity. In fact, the

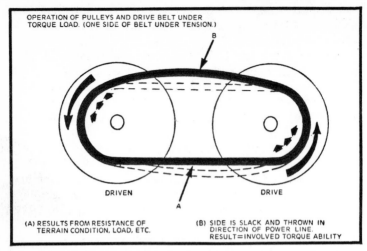

Fig. 9-7 V-belt response to load—giving some automatic torque multiplication (Courtesy Bombardier Ltd.)

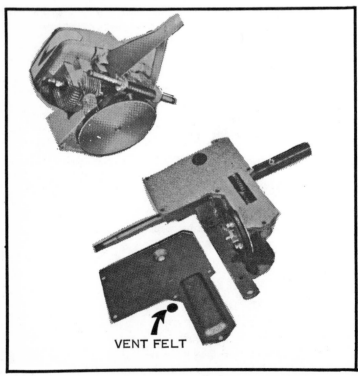

Fig. 9-8 Friction-drive disc and wheel (Courtesy OMC)

mechanism must be kept oil-free. A word of caution when storing this type of machine—the rubber wheel will flatten if the disc is left engaged.

A more popular friction drive is shown in Fig. 9-9 When the machine is put into gear, the drive rollers are lowered to come in contact with the wheels. With this style, roller slippage and forward speed is a function of pressure against the tires. Some drive rollers, favored on front-drive rotaries, are cogged for positive engagement. Better mowers employ needle bearings at the roller shaft, and bring with them the need for more meticulous periodic inspection. Cheaper machines get by with "self-lubricating" bushings; shaft end play in those units is critical since it determines the gear lash. On the machine pictured, the endplay is controlled by hardened steel thrustwashers between the right-hand roller and housing. Rollers are secured by setscrews or compression pins—or a combination of both. When knocking out the pins, use a ⅛ in. punch and support the shaft from below; otherwise, it may bend. Another point to remember about these machines is that the *belt lay (the way it's twisted) determines the direction of movement.* Install the belt wrong (backwards) and the machine will back over the operator!

CHAIN DRIVES

A properly set up chain drive will transmit 98% of the energy from the engine to the driven parts. Only 2% of the input is lost. This makes chains more efficient than gears or belts and, of course, friction drives. But this efficiency is rarely realized. To achieve it, the sprockets must be absolutely parallel, the chain links lubricated internally and externally, and the mesh between the chain rollers and sprocket teeth perfect. The relationship of chain-driven parts is shown in Fig. 9-10.

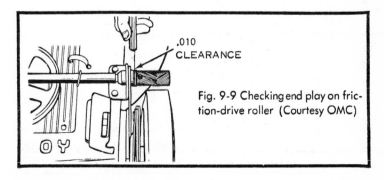

.010 CLEARANCE

Fig. 9-9 Checking end play on friction-drive roller (Courtesy OMC)

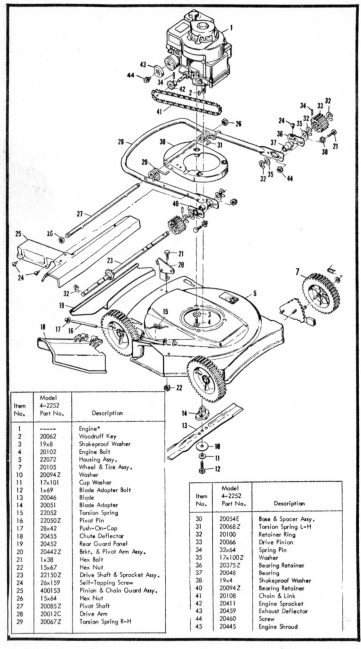

Item No.	Model 4-2252 Part No.	Description
1	-----	Engine*
2	20062	Woodruff Key
3	19x8	Shakeproof Washer
4	20102	Engine Bolt
5	22072	Housing Assy.
7	20105	Wheel & Tire Assy.
10	20094 Z	Washer
11	17x101	Cup Washer
12	1x69	Blade Adapter Bolt
13	20046	Blade
14	20051	Blade Adapter
15	22052	Torsion Spring
16	22050 Z	Pivot Pin
17	28x42	Push-On-Cap
18	20455	Chute Deflector
19	20452	Rear Guard Panel
20	20442 Z	Brkt. & Pivot Arm Assy.
21	1x38	Hex Bolt
22	15x67	Hex Nut
23	22150 Z	Drive Shaft & Sprocket Assy.
24	26x159	Self-Tapping Screw
25	400153	Pinion & Chain Guard Assy.
26	15x64	Hex Nut
27	20085 Z	Pivot Shaft
28	20012 C	Drive Arm
29	20067 Z	Torsion Spring R-H

Item No.	Model 4-2252 Part No.	Description
30	20054E	Base & Spacer Assy.
31	20068 Z	Torsion Spring L-H
32	20100	Retainer Ring
33	20066	Drive Pinion
34	32x64	Spring Pin
35	17x100 Z	Washer
36	20375 Z	Bearing Retainer
37	20048	Bearing
38	19x4	Shakeproof Washer
40	20094 Z	Bearing Retainer
41	20108	Chain & Link
42	20411	Engine Sprocket
43	20459	Exhaust Deflector
44	20460	Screw
45	20445	Engine Shroud

Fig. 9-10 Chain-driven rotary mower with positive-engagement rollers (Courtesy the Murray Ohio Mfg. Co.)

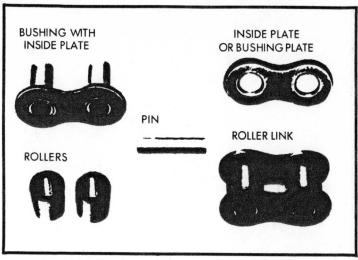

Fig. 9-11 Roller chain elements (Courtesy Daido Corp.)

Chains should be lubricated frequently; how frequently can be determined by inspecting the chain at the area of sideplate overlap. (See Fig. 9-11.) An oil-starved chain will have reddish (rust or burned oil) incrustations. When properly lubed, the chain will be black or blue.

There is a collection of folklore on the subject of chain lubrication which would fill several rooms of the Smithsonian. Most of this lore has been generated by motorcyclists in the U.S. and Great Britain. One school holds that the way to lubricate a chain is to boil it like an eel. The chain is removed from its sprockets, marinated in kerosene for a few days to dissolve rust and sludge, and then boiled in grease. The idea, of course, is to insure thorough internal lubrication.

While dirty or rusty chains should be removed and cleaned in solvent or carburetor cleaner, methods less drastic suffice for lubrication. Almost any lubricant—molybdenum disulfide, tungsten diselenide, graphite, lead, soaps—will do, if it is carried into the pins and bushings. Transmission oil (SAE 90) or chassis grease (cut with kerosene) are both effective. Light oils work, but must be applied frequently, since the oil is thrown off as the chain travels over the sprockets.

Chains should always have slack on the loose (nondriving) strand. The degree of looseness varies with the distance between sprocket centers, but should be at least ¼ in. of bow. Most reel mower jackshafts are adjusted by inserting shims

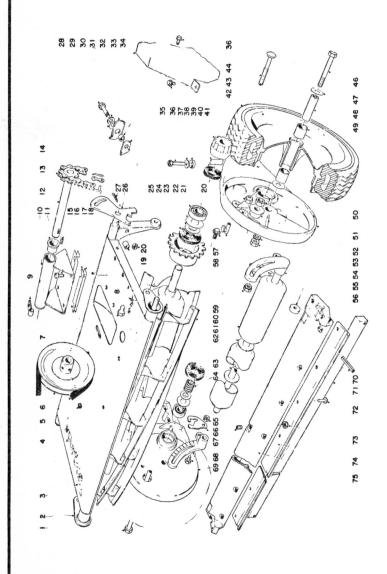

Fig. 9-12 Toro "Sportlawn" in exploded view (Courtesy the Toro Company)

MAIN FRAME ASSEMBLY

Ref. No.	Part No.	Description	No. Used 18"	No. Used 21"
1	5-4801	Tie Rod	●	
	5-4802	Tie Rod		1
2	5-1541	Motor Platform	1	
	5-1586	Motor Platform		1
3	17-8980	Decal	1	1
4	3245-1	Set Screw	2	2
5	271-17	V-Belt	1	1
6	5-1494	Pulley	1	1
7	3210-3	Capscrew	2	2
8	17-9110	Decal 21" (Hi-Cut)		1
	17-9000	Decal 21" (Lo-Cut)		1
	17-8990	Decal 18"	1	
9	2412-32	Clutch Control Cable Clip	1	1
10	3-7459	Countershaft Housing (Incl. Ref. 11)	1	
	3-7469	Countershaft Housing (Incl. Ref. 11)		1

Ref. No.	Part No.	Description	No. Used 18"	No. Used 21"
11	3-4066	Bushing	2	2
12	5-1551	Countershaft	1	
	5-1587	Countershaft		1
13	5-1552	Chain	1	1
14	2710-21	Connecting Link	1	1
15	3257-5	Woodruff Key	1	1
16	32113-18	Fil. Hd. Mach. Screw	4	4
17	5-1500	Shim	2	
	5-1588	Shim		2
18	5-1498	Shim	2	
19	5-1537	Spacer	2	2
20	32146-4	Locknut	4	4
21	251-30	Bearing	2	2
22	5-1520	Bearing Seal	2	2
23	5-1078	Thrust Washer	2	2
24	5-1518	Felt Washer	2	2
25	5-1572	Reel Sprocket	1	

Ref. No.	Part No.	Description	No. Used 18"	No. Used 20"
25	3-3065	Reel Sprocket		1
26	5-5250	Handle Bracket	2	
	5-5250	Handle Bracket (Low-Cut)		2
	5-5260	Handle Bracket (Hi-Cut)		2
27	3272-11	Cotter Pin	2	2
28	3290-213	Spring Washer	2	2
29	3210-15	Capscrew	2	2
30	3256-23	Flatwasher	2	2
31	5-1536	Stop Lever	2	2
32	5-1557	Spacer	2	2
33	5-1540	Clip L.H.	1	1
*	5-1538	Clip R.H.	1	1
34	3-5593	Chain Guard Top	1	1
35	5-1497	Adjusting Screw	4	4
36	32140-32	Self-Tapping Screw	2	2
37	5-1515	Upper Washer	4	4
38	5-1516	Lower Washer	4	4
39	5-1522	Felt Washer	2	2
40	2-3620	Ratchet Dog	2	2
41	2-2209	Pinion Kit (Incl. Right & Left Pinions & Dogs)	1	1
42	4-2580	Tire	2	2
	3-3668	Tire (Hi-Cut)		2

*Not Illustrated

Ref. No.	Part No.	Description	No. Used 18"	No. Used 21"
43	3-7389	Wheel w/Bushing (Incl. Ref. #49)	2	2
44	3-7253	Carriage Bolt	2	2
46	5-1513	Wheel Spindle Bolt	2	2
47	5-0744	Spindle Washer	4	4
48	5-1521	Wheel Spindle	2	2
49	256-32	Bushing	4	4
50	32146-2	Locknut	1	1
51	3-4789	Side Plate L.H. (Incl. Ref. #21 & 58)	1	1
52	3-4327	Tube End Spacer	2	2
53	3255-16	Lockwasher	2	2
54	3220-4	Jam Nut	2	2
55	2-3969	Roller Brkt. Set R.H. & L.H.	1	1
56	5-1517	Bed Bar Pivot	2	2
57	32140-16	Screw	3	3
58	301-2	Oil Cup	2	2
59	5-1505	Nut	2	2
60	4-7349	Reel Assembly	1	
	4-7329	Reel Assembly (Lo-Cut)		1
	4-7339	Reel Assembly (Hi-Cut)		1

Ref. No.	Part No.	Description	No. Used 18"	No. Used 20"
61	5-1553	Roller Section	3	
	5-1511	Roller Section		3
62	3256-6	Flatwasher	2	2
63	5-1554	Roller Shaft	1	
	5-1510	Roller Shaft		1
64	3-3081	Compression Spring	1	1
65	32146-1	Locknut	2	2
66	3-3856	Grass Catcher Hook	2	2
67	3245-9	Setscrew		1
68	329-3	Capscrew	2	2
69	3-7369	Side Plate Assembly R.H. (Incl. Ref. #21 & 58)	1	1
70	32121-12	Roll Pin		2
71	5-3498	Adjusting Bar		1
72	3290-57	Rivet	6	
	3290-46	Pan Hd. Rivet (Lo-Cut)		7
	3290-50	Pan Hd. Rivet (Hi-Cut)		7

Ref. No.	Part No.	Description	No. Used 18"	No. Used 21"
73	32127-24	Nut		1
74	5-1544	Bed Knife	1	
	5-1514	Bed Knife (Lo-Cut)		1
	3-3068	Bed Knife (Hi-Cut)		1
75	5-1543	Bed Bar	1	
*	3-7429	Bed Bar Assembly (Incl. Ref. #72 & 74)	1	
*	5-7739	Bed Bar Assembly (Lo-Cut) (Incl. Ref.#67 & 70-75)		1
*	5-7729	Bed Bar Assembly (Hi-Cut) (Incl. Ref #67 & 70-75)		1
*	4-1999	Bed Bar Assembly (Incl. Ref. #67, 70, 71, 73 & 75)		1
*	3-7449	Countershaft Assembly (Inc. Ref. #4, 6, 10, 12 & 15)		1

*Not Illustrated

337

(parts 17 and 18 in Fig. 9-12) to raise or lower the shaft. Other mowers may employ a chain tensioner or a sliding mount for one of the sprockets. Both methods are illustrated in Fig. 9-13. The latter—using elongated mounting holes—is standard for the rear wheels of riding mowers. When adjusting chains, take the bolts just a turn or so out, loosen both sides of the axle, and drive the wheels back. Keep the bolts tight enough to prevent the wheels from slipping and bouncing during adjustment; move both wheels evenly, keeping them parallel as you tighten the bolts to hold the adjustment.

Two tests are used to determine when chain wear dictates replacement. Perhaps the most meaningful is to remove the chain, count the links, and multiply by the pitch. The distance between the centers of each link—the pitch—in computation with the number of links, results in a nominal length-figure for the chain. With this information in mind, secure one end in a vise and pull it taut. If the actual length exceeds the nominal length by 3%, replacement is in order. The other test is to turn the chain on its side and, while holding it horizontally, examine it for more than ¼ in. deflection per running foot. More slack than that means that the side plates have worn excessively.

Chains follow the National American Standards Institute coding. The first digit indicates the pitch of the chain in eighths of an inch. The second describes the chain construction; 0 is standard, 1 is lightweight. A 5 is a chain without rollers. Thus, a 41 chain has a pitch of ½ in. ($^4/_8$), and is intended for light-duty applications. This type of chain is used on many of the cheaper mowers and, where clearance allows, can be replaced by a sturdier No. 40.

Lawnmower chains have a master (breakable) link that is released by flexing the chain toward the link as shown in Fig. 9-14. The link is identified by its ovoid shape and should be on the accessible side of the sprockets. Another style of master link is shown as H in Fig. 9-15. These links are broken by prying the spring lock ends up and over the pin. In general, three-element links are stronger than the two-element bicycle types and should be used whenever the space is available for them. They cannot, for example, be used on the *derailleur*-like two-speed mechanism found on some Cooper rotaries.

GEAR DRIVES

Figure 9-16 depicts a spur-gear drive used on many light rotary mowers. These gears run in bushings in a pot-metal

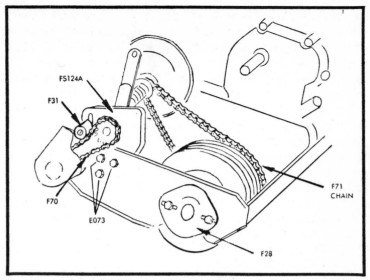

Fig. 9-13 Cooper "Klipper-Trim" chain adjustment is by an idler (F 31) and by elongated bolt holes at the wheels (F 28) (Courtesy Cooper Mfg. Co.)

housing. The bushings require examination for wear occasionally, and particular attention should be paid to the gear shafts. Earlier remarks about the low quality of lawnmower shafting apply especially to these boxes. However, the OMC unit shown seems generally superior to the others and does not require modification. The gears are secured by roll pins and, in most cases, are only surface-hardened; rapid wear becomes a problem once the surface is broken. Always replace gears in matched (new or used together) pairs and pack the box with grease.

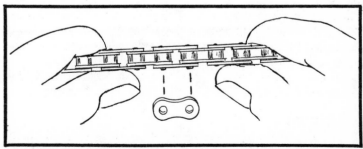

Fig. 9-14 Master link removal (Courtesy Cooper Mfg. Co.)

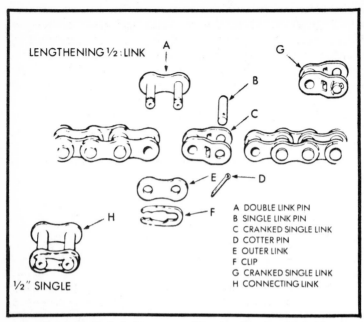

LENGTHENING ½ : LINK

A DOUBLE LINK PIN
B SINGLE LINK PIN
C CRANKED SINGLE LINK
D COTTER PIN
E OUTER LINK
F CLIP
G CRANKED SINGLE LINK
H CONNECTING LINK

½" SINGLE

Fig. 9-15 Chain link variations (Courtesy Bombardier Ltd.)

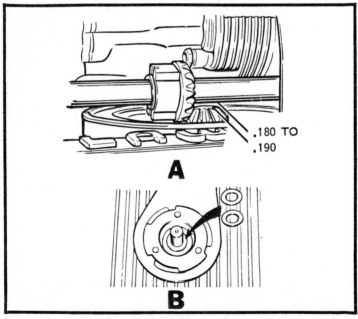

.180 TO
.190

A

B

Fig. 9-16 A. Spur gear set. B. Spacers to fix lash (Courtesy OMC)

Figure 9-17 illustrates an exploded view of the Toro front-wheel drive used on their rotary mowers. It is surely one of the more conceptually sophisticated and better-constructed propulsion mechanisms ever fitted to lawnmowers. Speed reduction is in three stages—the belt sheaves, the worm and helical gears in the box, and the pinion gears at the wheels. The helical gear is engaged through dog ramps (part 38) for smooth clutch action. Shaft bushings feature seals, and the worm is supported on a Conrad ball bearing that can accept side and thrust loads. This mechanism requires very little maintenance aside from periodic lubrication.

TRANSMISSIONS

Selective-gear transmissions are generally the province of riding mowers, although you will encounter several varieties on some of the more elaborate rotaries. Pincor reels, long since forgotten, had two-speed boxes.

Since engines are pretty well standardized (the difference between the three major brands are on the order of the differences between the major automotive offerings), riding mower quality generally shows up in the running gear. The transmission is most vulnerable to cost-saving shortcuts. The cheapest mowers get along with one forward speed, neutral, and reverse. As the price goes up, so does the number of speeds. The quality of the box varies with the steel and heat treatment, the type of bearings (antifriction bearings are preferable to bushings), and the gear selector mechanism. Sliding gear transmissions are aptly described as "crash boxes," since the gears must move along their shafts and engage at the teeth. Constant-mesh transmissions shift easier and quicker since the main shaft does not have to match the layshaft speed before engagement.

There are some general service and troubleshooting principles which should be mentioned before discussing the various selective gear transmissions:

- Try the shift mechanism. Failure to shift may be the fault of the clutch or (in a few cases) the part of the selector mechanism outside of the box.
- Oil leaks are usually evidence of bad bearings. Some of the cheaper boxes do not have seals, but depend entirely on the fit between the shaft and bushing; when seals fail, the fault is usually in the bearing or shaft.

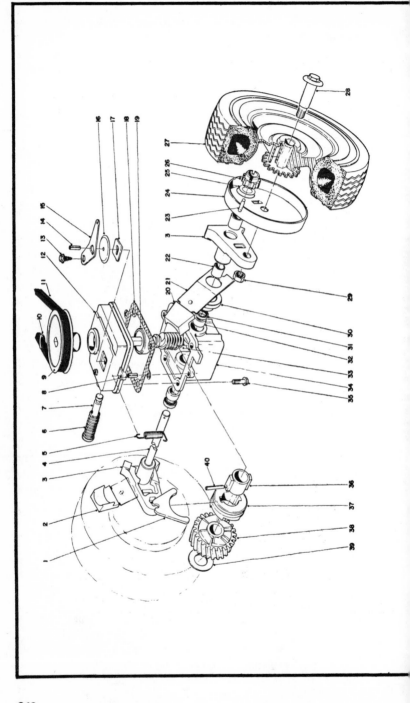

Ref. No.	Part No.	Description	No. Used
1	5-3247	Shifter Fork	1
2	6-6220	Knob Arm	1
3	18-9710	Pivot Arm (Incl. Ref. No. 22)	2
4	18-9730	Drive Shaft	1
5	17-4490	Spring Tension	1
6	3-0731	Spring	1
7	5-3248	Shifter Rod	1
8	32121-1	Roll Pin	2
9	18-9760	Pulley	1
10	3290-346	Push-on Retainer	1
11	5-4505	V-Belt	1
12	5-3283	Cover	1
13	32140-64	Self Tapping Screw	1
14	32121-23	Roll Pin	1
15	5-3252	Shifter Arm	1
16	3-2615	Dust Cover	1
17	3-0681	Latch Plate	1
18	3-0734	Gasket	1
19	251-144	Ball Bearing	1
20	6-6230	Knob Arm	1
21	3-2640	Spring Arm	2
22	256-1	Bushing	4
23	3284-3	Groove Pin	2
24	18-9820	Wheel Cover	2
25	18-9810	Thrust Washer	2
26	18-9690	Pinion Gear	2
27	11-9470	Wheel and Tire Assembly	2
28	3-2652	Wheel Bolt	2
29	32146-12	Locknut	2
30	5-0745	End Cap	2
31	253-86	Oil Seal	2
32	256-110	Bushing	2
33	18-9700	Worm Shaft	1
34	5-4459	Gear Case (Incl. Ref. No. 31 & 32)	1
35	11-9760	Screw	4
36	3-0721	Sleeve	1
37	19-5660	Clutch Jaw	1
38	18-9740	Helical Gear	1
39	3-0724	Thrust Washer	1
40	32121-88	Coil Pin	1
*	18-9780	Complete Gear Case Assembly	1

*Not Illustrated

Fig. 9-17 Worm and helical gear (Courtesy The Toro Company)

- Scrutinize the oil. Metal particles generally mean trouble—although they can be generated by careless gear clashing.
- If it is necessary to dismantle the box, remove it from the mower. Notice how the workings operate together as you disassemble the box. Understanding the principles involved will make troubleshooting and assembly easier. It is wise to lay out the parts on clean newspaper from left to right (as you would read a page). Pay particular attention to the location of the thrust washers, as they determine the gear lash and protect the soft metal case from contact with the gears.
- Examine the gears for missing, chipped, or worn teeth. Sliding gearboxes have chamfered teeth to make engagement easier. Beveled edges cause deeper tooth wear, but it won't be critical until it approaches half the tooth width.
- Replace bearings as required. Bushings are not finish-reamed since the working clearances conform to broad tolerances. Figure 9-18 shows one method of removing bushings from closed-end bores. The diameter of the bolt varies depending on the application. Another method, based on the principles of a hydraulic ram, was outlined in the previous chapter. Needle bearings should use oiled newspaper as the hydraulic medium rather than grease. It may help to apply heat to the case. Shaft bearings are removed and installed in the same way as major antifriction parts would be.
- Inspect shafts for trueness. This step is very important when dealing with a bargain-counter box. (Shafts can be straightened with an arbor press.)
- Be wary of those transmissions that have spring-loaded detent balls; they can easily pop into oblivion.
- Inspect the shifting fork for bends and extreme wear. In most instances, shifting forks can be repaired by welding and straightening.

Figure 9-19 shows an all-direct (power passes through the lay shaft regardless of the gear selected). Peerless three-speed box; the reverse gear idler is not shown. Gear shift selection is made by moving the gears fore and aft on the input shaft with the mechanism shown in Fig. 9-20. The 200 series is a constant-mesh design in which the gears spin freely until

344

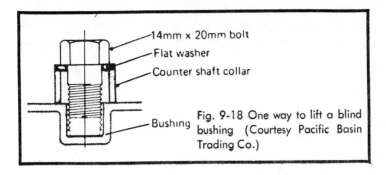

14mm x 20mm bolt
Flat washer
Counter shaft collar

Bushing

Fig. 9-18 One way to lift a blind bushing (Courtesy Pacific Basin Trading Co.)

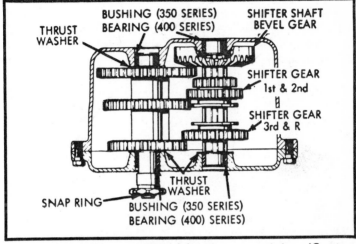

BUSHING (350 SERIES)
BEARING (400 SERIES)

SHIFTER SHAFT BEVEL GEAR

THRUST WASHER

SHIFTER GEAR 1st & 2nd

SHIFTER GEAR 3rd & R

THRUST WASHER

SNAP RING

BUSHING (350 SERIES)
BEARING (400) SERIES

Fig. 9-19 Peerless Series 350 sliding-gear transmission (Courtesy Tecumseh Products Co.)

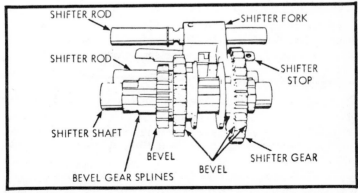

SHIFTER ROD

SHIFTER FORK

SHIFTER ROD

SHIFTER STOP

SHIFTER SHAFT

BEVEL

BEVEL GEAR SPLINES

BEVEL

SHIFTER GEAR

Fig. 9-20 Shifter mechanism for the 350 transmission (Courtesy Tecumseh Products Co.)

345

locked by the shifter lug which is keyed to the input shaft (Fig. 9-21).

A much more elaborate transmission (Fig. 9-22) is fitted to the International Cadet 55, 75, and 95 series mowers and the Murray tractors. Power enters from below through the input shaft (17). The bevel gear (23) is keyed to the main or drive shaft (32). Power can, depending upon the position of the shift fork (59) and clutches (8), go to the output or lay shaft through the chain (60) for reverse. Three gears (7), (12), and (24), provide the forward speeds. Braking is provided at the outboard disc (47).

To disassemble, remove the top cover (37) and lift out the drive and output shafts together. Slip the reverse chain from its sprockets. The output shaft is disassembled at the brake end; the key component is the retaining ring (4). The procedure is similar for the drive shaft. To dismantle the shift mechanism, first remove the four screws holding the shift lever to the cover (43), lift the lever clear, and remove the insert (42) and wave washer (41). Remove the safety start switch (58) and spring (57). The easiest way to remove the pin (55), ball (54), springs (40), and poppet balls (39), is with a magnet. Assembly is in the reverse order. The most critical job is to shim the backlash between the bevel pinion (21) and the bevel (23). There are various ways to do this, but the only *approved* method is to use a dial indicator to obtain $0.001 - 0.007$ in. play between pairs of engaged teeth. The other dimensions can be checked with a feeler gage (Fig. 9-23). The gap between the flange bearing (5) and the sprocket (25) should be 0.001 to 0.01 in.—the same clearance applies to the inner edge of the output sprocket (2) and the outer side of the lower housing. These dimensions are brought into congruity with shims. They must be verified after long use or if major parts such as bearings and gear sets have been changed.

DIFFERENTIALS

Figure 9-24 is a phantom view of a typical heavy-tractor differential. During straight-ahead operation, both axle shafts turn with the differential case which is riveted to the ring gear. In a turn, the inside axle slows and power goes through the spider gears to the outside (least loaded) axle. The spider gears differentiate (choose) the least resistive path to deliver their torque; all power is transmitted to the free wheel. On the other hand, if both wheels are loaded equally, the spiders favor

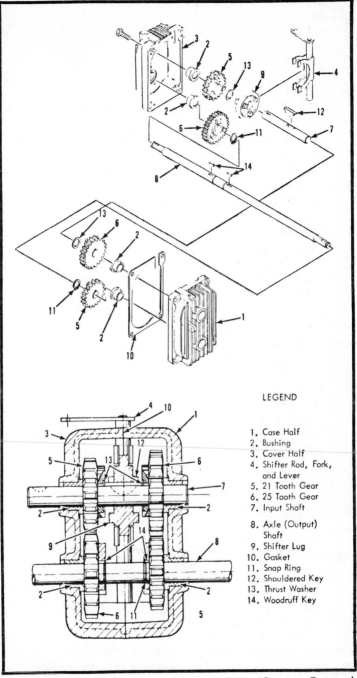

LEGEND

1. Case Half
2. Bushing
3. Cover Half
4. Shifter Rod, Fork, and Lever
5. 21 Tooth Gear
6. 25 Tooth Gear
7. Input Shaft

8. Axle (Output) Shaft
9. Shifter Lug
10. Gasket
11. Snap Ring
12. Shouldered Key
13. Thrust Washer
14. Woodruff Key

Fig. 9-21 Peerless 200 constantmesh transmission (Courtesy Tecumseh Products Co.)

347

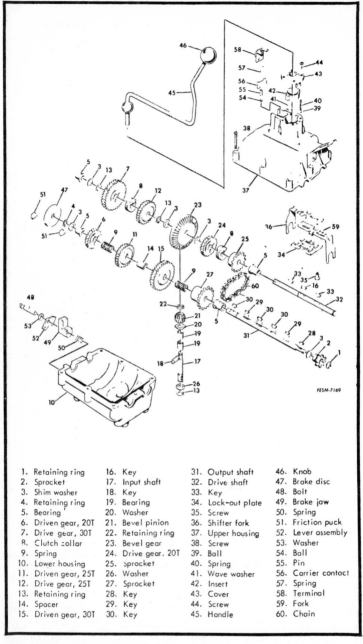

1. Retaining ring	16. Key	31. Output shaft	46. Knob
2. Sprocket	17. Input shaft	32. Drive shaft	47. Brake disc
3. Shim washer	18. Key	33. Key	48. Bolt
4. Retaining ring	19. Bearing	34. Lock-out plate	49. Brake jaw
5. Bearing	20. Washer	35. Screw	50. Spring
6. Driven gear, 20T	21. Bevel pinion	36. Shifter fork	51. Friction puck
7. Drive gear, 30T	22. Retaining ring	37. Upper housing	52. Lever assembly
8. Clutch collar	23. Bevel gear	38. Screw	53. Washer
9. Spring	24. Drive gear, 20T	39. Ball	54. Ball
10. Lower housing	25. Sprocket	40. Spring	55. Pin
11. Driven gear, 25T	26. Washer	41. Wave washer	56. Carrier contact
12. Drive gear, 25T	27. Sprocket	42. Insert	57. Spring
13. Retaining ring	28. Key	43. Cover	58. Terminal
14. Spacer	29. Key	44. Screw	59. Fork
15. Driven gear, 30T	30. Key	45. Handle	60. Chain

Fig. 9-22 Three-speed reversing transmission (Courtesy International Harvester)

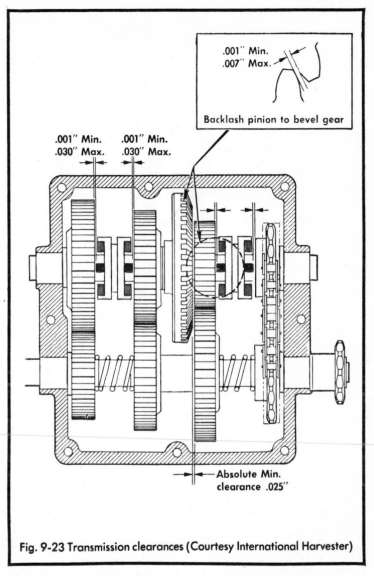

Fig. 9-23 Transmission clearances (Courtesy International Harvester)

neither. The differential is one of those immensely clever devices invented in a totally abstract frame of reference by a sixteenth-century French mathematician, in the course of his studies hundreds of years before a torque-splitting gear train was needed. Posed as an engineering problem today, one would be hard-pressed to come up with a solution of equally elegant simplicity. Figure 9-25 illustrates the device's place in

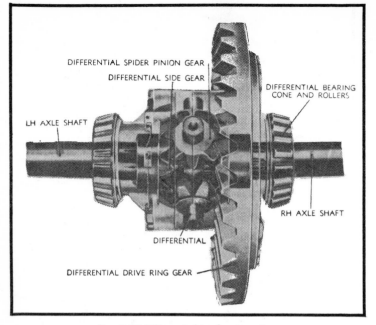

Fig. 9-24 Differential in phantom view.

the contemporary lawnmower. A modification of the basic concept—the limited-slip differential—is exemplified by the *Duo-Trak* (from Illinois Tool Works) found on many riding mowers (Fig. 9-26), in conjunction with a Peerless 2300 or 2400 transmission. These units are combined in a common case and, thus, are referred to as *transaxles*.

Points of note in these units are as follows:

- Some models are off the center line of the frame and have axles of different lengths.
- The brake springs hold the pinions in place, and can be removed with a hefty pair of snapring pliers.
- Assemble by aligning the cores so that the pockets of one core do not match the pockets of the second.
- Install five pinion gears in mesh with the side gear—turn the unit over and install the other five to mesh with the pinions already in place.
- The brake spring bottoms on the side gear. It should be in contact with most of the pinions.
- The through-bolts should be torqued to 7 – 10 ft-lb.

Note: the Duo-Trak is not a locked differential—if one wheel is clear of the ground, it will spin.

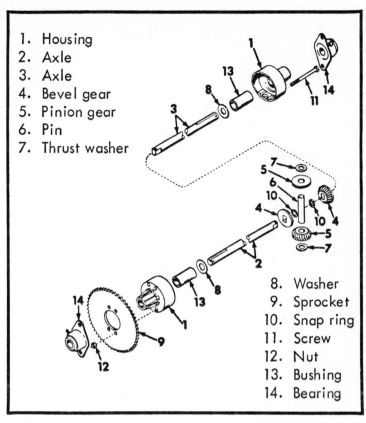

1. Housing
2. Axle
3. Axle
4. Bevel gear
5. Pinion gear
6. Pin
7. Thrust washer

8. Washer
9. Sprocket
10. Snap ring
11. Screw
12. Nut
13. Bushing
14. Bearing

Fig. 9-25 Riding mower differential (Courtesy International Harvester)

CLUTCHES

Clutches can be very simple. The most frequently encountered are idler pulleys located between the sheave centers on V-belt drives (Fig. 9-27). Depending upon the application, available space, and designer preferences, the idler may be on the inside or the outside circumference of the belt. Merely varying the belt tension is not enough to disengage the belt from its grooves—the belt guide must be in rubbing contact with the outside surface of the slack side of the belt to nudge it up and against the rotating sheave. One or more contact points on the taut side act as fulcrums.

Next to the belt (which slips until it matches sheave speed during engagement—and bumps and snatches during disengagement), the idler pulley bearing is most vulnerable to

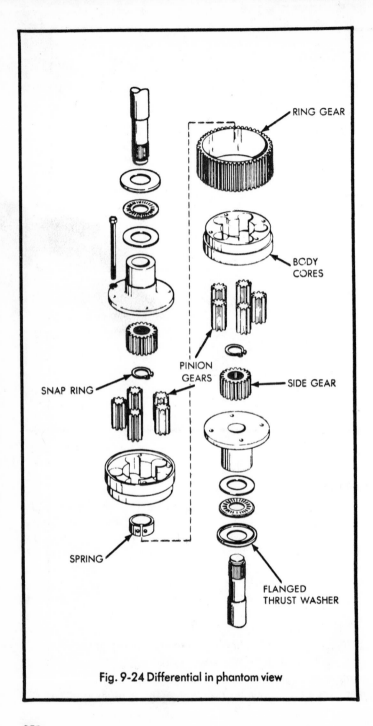

RING GEAR

BODY CORES

PINION GEARS

SNAP RING

SIDE GEAR

SPRING

FLANGED THRUST WASHER

Fig. 9-24 Differential in phantom view

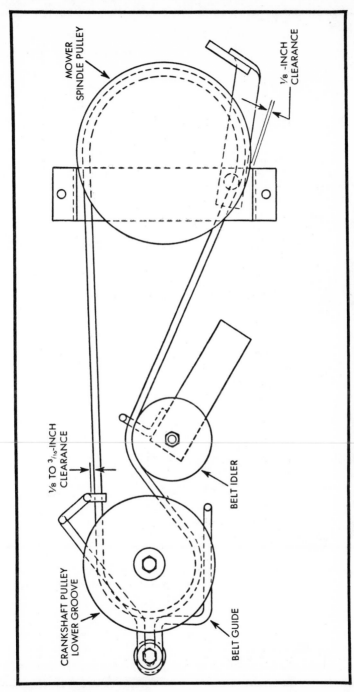

MOWER SPINDLE PULLEY

⅛-INCH CLEARANCE

⅛ TO ³⁄₁₆-INCH CLEARANCE

CRANKSHAFT PULLEY LOWER GROOVE

BELT IDLER

BELT GUIDE

Fig. 9-27 V-belt idler clutch (Courtesy Tecumseh Products Co.)

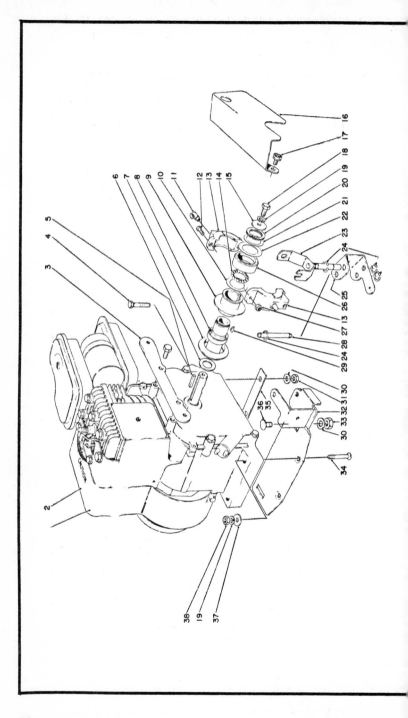

ENGINE ASSEMBLY

Ref. No.	Part No.	Description	No. Used 18"	No. Used 21"
1	221-448	Engine	1	1
2	12-6590	Decal Caution	1	1
3	3-7658	Bracket	1	1
4	3210-15	Capscrew	2	2
5	5-1077	Square Key	1	1
6	5-1574	Washer	1	1
7	5-1558	Clutch Body	1	1
8	5-1546	Idler Bushing	1	1
9	5-1545	Sliding Flange	1	1
10	5-1548	Thrust Washer	1	1
11	5-1555	Oiler	1	1
12	3250-4	Machine Screw	2	2
13	3-3517	Slip Ring	2	2
14	255-5	Steel Balls	18	18
15	4-8620	Washer	1	1
16	3-7661	Belt Guard	1	1
17	32140-39	Machine Screw	2	2
18	329-4	Capscrew	1	1
19	3253-3	Lockwasher	3	3
20	32120-14	Snap Ring	1	1
21	5-1550	Check Washer	1	1
22	5-1226	Slip Ring Washer	1	1
23	5-4291	Shift Fork	1	1
24	32120-35	Snap Ring	4	4
25	5-4293	Shift Fork Bracket	1	1
26	5-1547	Shift Collar	2	2
27	3296-1	Nut	1	1
28	3-4031	Pivot Shaft	1	1
29	3257-1	Woodruff Key	1	1
30	3256-23	Flatwasher	4	4
31	32146-4	Nut	2	2
32	3-5780	Clutch Mounting Plate	1	1
33	32146-5	Nut	2	2
34	3289-10	Flat Head Screw	1	2
35	5-4294	Shim	2	2
36	3230-1	Carriage Bolt	2	2
37	3256-2	Flat Washer	2	2
38	3219-1	Nut	2	2
*	2-3029	Clutch Assembly (Incl. Ref. No's. 7-15 and 18-22, 26 & 29)	1	1
*	2-6739	Loose Parts Kit	1	
*	2-6769	Loose Parts Kit (Hi-cut)		1
*	4-7249	Loose Parts Kit (Lo-cut)		1

*Not Illustrated

Fig. 9-28 Toro split sheave clutch (Courtesy the Toro Company)

wear. It undergoes rapid acceleration when the machine is put into gear and turns at rather high speeds. A bushing is inadequate, although it is all that some manufacturers provide. Ideally, the idler should turn on a ball bearing with a greased fitting. Prelubed bearings seem to fail early in this harsh environment, but last longer than bearings which are neglected. Few owners of lawnmowers understand the importance of preventive maintenance, or want to, when it entails work on the underside of a riding mower.

Other service procedures are to adjust the throw of the idler arm for belt stretch, to check the running clearance of the belt guards (the dimensions in Fig. 9-27 are good approximations for all machines), and to inspect the idler pulley for wear and for sharp edges which would cut the belt. The idler must be parallel to the belt for best results; some bending of the linkage may be required to maintain this condition.

SHEAVE CLUTCHES

A much more sophisticated clutch is shown in the next illustration (Fig. 9-28). Used on Toro reel mowers for many years, it treats belts almost tenderly. The sheave is split into a sliding half and into a stationary flange (or in Toro's nomenclature, a "clutch body"). These parts are illustrated as 9 and 7 in the illustration. The sliprings (13) move the sheave halves together and apart in response to movement of the shift fork (23). In neutral the belt rides on the idler bushing (8).

The only routine maintenance required is to oil the clutch at the lidded port (11). Failure to engage usually means maladjustment of the control cable; failure to remain engaged can sometimes be corrected by loading the cable to keep pressure on the yoke, but is best repaired by rebuilding the unit. Toro offers small parts kits for this purpose. It is advisable when removing the clutch from the engine to place it on a clean bench so you can keep track of the parts sequence and control the rain of ball bearings.

Automatic clutches are usually centrifugal devices which engage at a predetermined speed. Some wonderful and ingenious designs have been tried; a few used oil as the medium and doubled as torque converters. But the simplest, and by all means the most popular, is the centrifugal clutch. It consists of two or more weighted clutch shoes, a drum, a spider to locate the shoes, return springs, and an idler bearing.

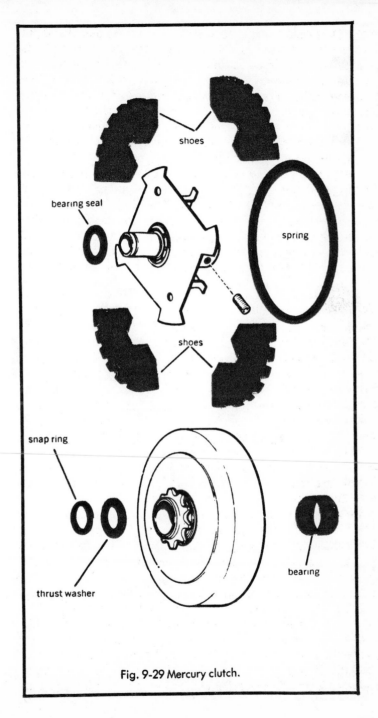

Fig. 9-29 Mercury clutch.

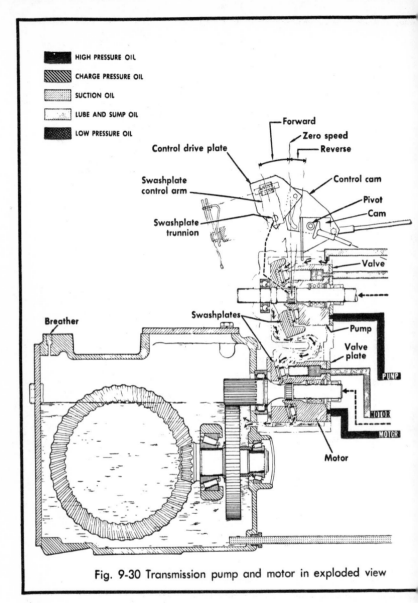

HIGH PRESSURE OIL

CHARGE PRESSURE OIL

SUCTION OIL

LUBE AND SUMP OIL

LOW PRESSURE OIL

Forward

Zero speed

Reverse

Control drive plate

Control cam

Swashplate control arm

Pivot

Cam

Swashplate trunnion

Valve

Breather

Swashplates

Pump

Valve plate

PUMP

MOTOR

MOTOR

Motor

Fig. 9-30 Transmission pump and motor in exploded view

Mercury has built so many of these clutches that the trade name is often applied indiscriminately to all centrifugal clutches—regardless of manufacturer. The clutch in Fig. 9-29 is a genuine Mercury.

The only point needing lubrication (and a minuscule amount at that) is the drum bearing. No oil should ever be

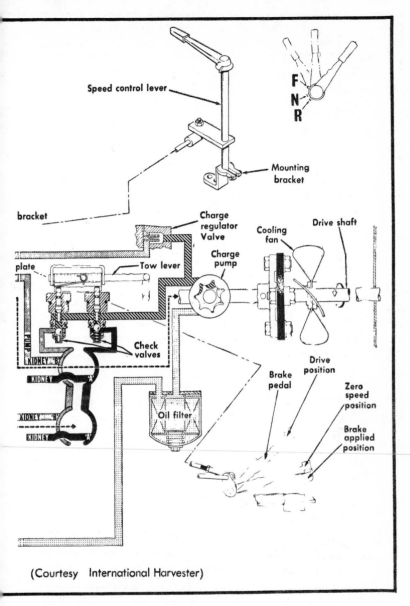

allowed to get on the clutch shoes, nor should the shoes be cleaned with solvents. Since the linings usually contain a high percentage of asbestos, it is wise to hose the clutch with water prior to full disassembly to settle the dust.

The drum bearing supports the drum while the clutch is disengaged, separating the spinning crankshaft from the idle drum. During engagement, it absorbs running loads. The shoes merely grip the inner diameter of the drum and do not contribute to its support. The outboard sprocket or sheave adds to the bearing's task by concentrating loads on one end; consequently, the rate of wear is higher than one would suppose from looking at its generous dimensions. When the drum shows wobble or erratic movement in neutral, drive the bearing out and use a replacement. If possible, ream the part to a smooth finish.

As the friction material on the shoes wears, the speed at which engagement takes place becomes progressively higher. Most of these clutches are designed to engage on the shy side of 2000 rpm. Extreme wear results in slippage under load. Early engagement or failure to entirely disengage is usually caused by heat-frazzled springs, although an out-of-round drum could be a contributing factor.

INFINITELY VARIABLE TRANSMISSIONS

Some of the larger tractors employ hydrostatic transmissions. A "hydro" has become the ultimate status symbol in the world of grass-cutting. No other device gives such precise control over an internal combustion engine. The heart of the transmission is a pump paired to a hydraulic motor. The pump is driven by the engine and is controllable in infinite increments in terms of fluid flow rate and pressure. The rate of flow determines the speed of the motor and wheels—the pressure determines the magnitude of force generated by the motor. In addition, the direction of flow is controllable so that the machine can be reversed.

Control is achieved by means of a swash plate which can be tilted off vertical. The plate bears against nine pistons which comprise the pumping elements (Fig. 9-30). The angle of the swash plate, which determines the displacement of the pump, is controlled by the operator. Without a change in angle, no reciprocation occurs and the machine is in neutral. The connection between the pump and its motor is through a mass of porting known as the kidneys. With the drive shaft rotating clockwise (as viewed from the drive end) and the swash plate tilted to the rear, kidney B will be the inlet and kidney A will be the outlet. As the pump cylinder block rotates past kidney B, fluid is drawn into the piston bores; then, as rotation continues, fluid is expelled into kidney A by the action of the pistons. The greater the angle of the swash plate, the greater the displacement, causing more fluid to be delivered on each revolution of the cylinder block. When the swash plate is tilted to the front, the flow reverses, and kidney A becomes the inlet with kidney B becoming the outlet—the machine reverses.

The motor is an axial-piston fixed-displacement design. The cylinder block is splined to the output shaft so both parts rotate together. As fluid under pressure is introduced by the pump, the pistons are forced against the swash plate. Since the plate is tilted at a fixed angle, the pistons are moved on the plate by cam action and rotate the motor cylinder block. Fluid is expelled during cylinder rotation.

A charge pump furnishes fluid to make up for leakage and provide circulation for cooling. Any surplus output may be diverted to operate hydraulic accessories.

Service other than checking the oil level and hose connections is best left to factory-trained mechanics. The transmission appears to be conceptually simple and obvious in function, but the parts are machined and finished to extreme tolerances and warrant special service procedures.

The big green Porter-Cable machines and other "riders" employ a less sophisticated variable-speed transmission in the form of a torque converter. It consists of a pair of split sheaves and a belt—the sheave opening and closing according to load and engine speed. This type of transmission is less flexible than a hydrostatic drive, since overall torque multiplication is limited to a ratio of approximately 3 to 1. The Colt Industries unit illustrated gives an initial ratio of 3.1 to 1; that is, the drive sheave turns slightly more than three revolutions for each revolution of the driven member. The final ratio is 0.95 to 1. Figure 9-31 shows the complementary way in which the effective ratios of both sheaves are changed. As one increases, the other decreases, keeping belt tension more or less constant.

The drive sheave is speed-sensitive and works on the same centrifugal principle described in an earlier discussion about clutches. But instead of employing shoes which grip the inner surface of the drum, the drive sheave uses weights which cam the sheave together as they are flung outwards (Fig. 9-32). At idle, the belt rides on the bottom of the sheave without contacting the sides. A free-spinning bushing absorbs all of the crankshaft torque when the machine is in neutral. Off idle, the weights move outward and bring the sheave halves closer together. At some point dictated by weight geometry, the belt

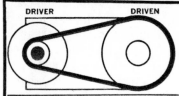

NEUTRAL POSITION

There is no engagement while the engine is idling.

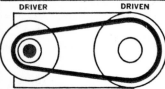

LOW RANGE

As the engine speed is increased the Driver pulley starts to close which engages the drive belt. This in turn drives the Driven pulley at its most powerful ratio.

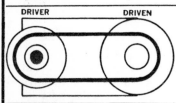

MIDDLE RANGE

As speed increases the Driver pulley continues to close. The belt is forced out further on the diameter of the Driver. This increases the acceleration of the unit. The drive ratios are infinite between high and low speeds.

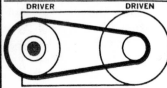

HIGH RANGE

At top speed the Driven pulley flanges are wide open and the Driver flanges are closed. This provides the highest speed of the vehicle (overdrive).

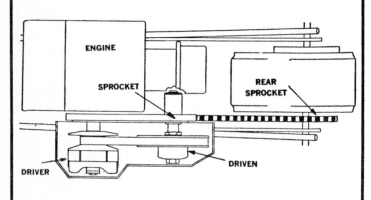

Fig. 9-31 Torque converter operation.

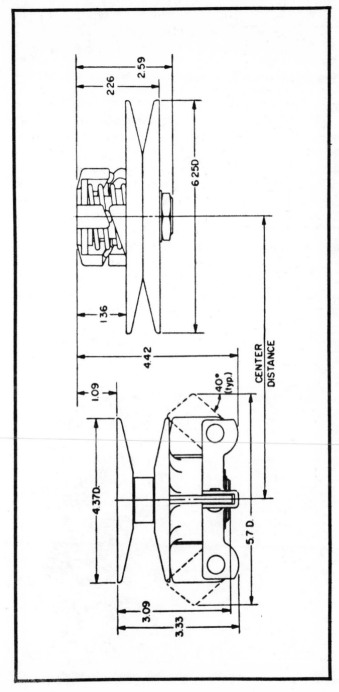

Fig. 9-32 An engineering drawing of the Colt 850 series converter (Courtesy Colt Industries)

is gripped by the sheave halves and transmits torque. The cut-in point is tailored to engine characteristics, and occurs early in the speed curve for industrial engines fitted to riding lawnmowers. Sheave movement is progressive and in direct response to the position of the weights which, in turn, is a function of shaft speed. The closer the sheaves move together, the greater the effective sheave diameter. The end of travel coincides with a falloff in torque at 3200−3400 rpm. A further increase in effective diameter would only overgear the engine.

The driven sheave is load-sensitive; under light loads it compresses to its full diameter because of the coil spring. As loads increase, the cam-side half slows relative to the fixed side. This difference in velocity is translated to movement on the cam which allows the sheave to open, reducing the effective diameter.

Normally, the only maintenance torque converters require is periodic belt inspection and changes. Belts take a severe beating as they slide up and down on the sheaves. The drive sheave may require cleaning to restore its action, and the driven member needs some lubrication at the cam ramp. If you take one apart (hardly ever necessary,) be careful when you release the spring. It can blow the sheaves apart with great force.

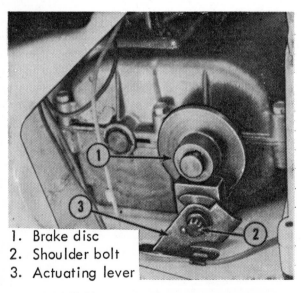

1. Brake disc
2. Shoulder bolt
3. Actuating lever

Fig. 9-33 Disc brake (Courtesy International Harvester)

BRAKES

Riding mowers are often equipped with blade brakes as a safety feature. This adds complexity to the machine, since in addition to a brake mechanism there must be provision to disengage the blade from the engine. But the complexity is worth it. The brake consists of a shoe which is levered against the blade drive member in coordination with clutch disengagement. The linkage is adjustable to compensate for wear. Shoes can be replaced with parts available at your dealer or, in a pinch, linings from well-furnished automobile brake shop can be bonded to them.

Riding mower wheel brakes are usually found on the transmission and may be in the form of a puck and disc (Fig. 9-33) or a drum and externally contracting linings (Fig. 9-34). Service entails checking the brake elements for clearance in the disengaged position and replacing the friction material when worn.

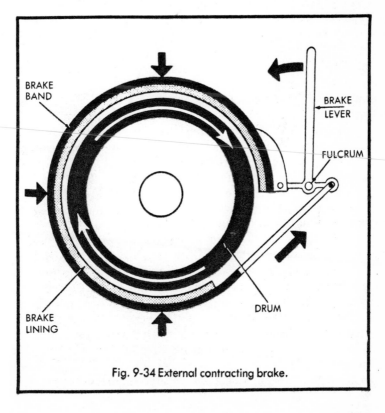

Fig. 9-34 External contracting brake.

Chapter 10

The KM3 Rotary

The Wankel, or rotary—or if you follow Curtiss-Wright's and OMC's nomenclature—the rotary combustion engine, was developed by Dr. Felix Wankel and the NSU engineering staff in 1954. Ten years later the first rotary engine entered the market under the deck of an NSU Spider. Since then, almost 30 firms have received manufacturing rights from NSU. Wankel engines are used in three makes of automobile and have been built for aircraft and multifuel configurations.

Fichtel & Sachs is the preeminent builder of rotary engines in the 20-horsepower-and-under class. Rotaries account for a hefty slice of F&S production and currently come off the line in Schwienfurt at a rate of 20,000 a month. The outstanding feature of these engines is their almost dead smoothness which is a particular advantage in light vehicle applications. As Helmut Keller has pointed out in an SAE paper, F&S rotaries are ideal for rail grinding, rock drilling, and circular saw applications. The absence of vibration gives extended life to the cutting tools and produces a uniform finish. These engines are able to materially increase the life of small hulls and glider frames. In 1973 F&S released the KM 3, designed specifically for lawnmower use.

The KM 3 (Fig. 10-1) is a fan-cooled single-rotor engine which develops 4 hp (SAE) at 4500 rpm. To insure acceptable seal life and to reduce blade tip speeds, the engine is governor-limited to a speed between 2300 and 2700 rpm. Dry weight is fractionally less than 22 lb. Fuel economy is superior to 2-cycles in the same class, but inferior to 4-cycles. Exact figures are not available for the KM 3, but on the basis of other F&S rotaries, comsumption can be expected to fall between 0.8 and 0.6 lbs/hp/hr. The cost is higher than for a comparative reciprocating engine because of developmental expenses,

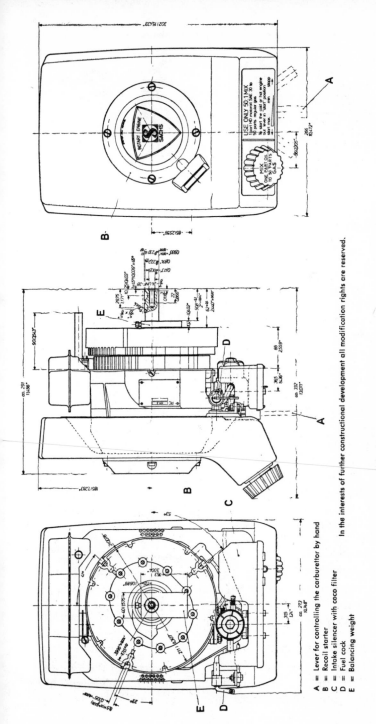

USE ONLY 50:1 MIX
Use HD mineral SAE 30 to
50 parts regular gas

To start the cold or hot engine
put lever in "start" position.

ROTARY ENGINE
SACHS

MIX
ONE PART OIL
TO 50 PARTS
GAS

In the interests of further constructional development all modification rights are reserved.

Fig. 10-1 The KM 3 lawnmower engine (Courtesy Fichtel & Sachs)

A = Lever for controlling the carburettor by hand
B = Recoil starter
C = Intake silencer with coco filter
D = Fuel cock
E = Balancing weight

367

1. Intake.

Fuel/air mixture is drawn into combustion chamber by revolving rotor through intake port (upper left). No valves or valve-operating mechanism needed.

2. Compression.

As rotor continues revolving, it reduces space in chamber containing fuel and air. This compresses mixture.

3. Ignition.

Fuel/air mixture now fully compressed. Leading sparkplug fires. A split-second later, following plug fires to assure complete combustion.

4. Exhaust.

Exploding mixture drives rotor, providing power. Rotor then expels gases through exhaust port.

Fig. 10-2A How the rotary engine works.

1. Intake

2. Compression.

3. Ignition.

4. Exhaust.

Fig. 10-2B How the piston engine works (Courtesy Toyo Kogyo)

relative low production volume, and because engines which are sold in North America are franchised through Curtiss-Wright at a premium. In Europe, F&S rotaries are competitive in the 10−20 hp range, although there is a cost penalty for the smaller sizes.

These engines are very sensitive to variations in oil types used. Automotive and marine oils have been known to cause excessive carbon deposits in a few hours of operation. The F&S firm markets an oil formulated for rotary engines which works better than other brands. Regular gasoline is mixed with the oil at a 50:1 ratio prior to filling the tank.

OPERATING PRINCIPLES

The rotary engine (Fig. 10-2A) operates on the familiar Otto working cycle (Fig. 10-2B) but events are telescoped in a manner somewhat similar to a 2-cycle engine. The rotor sweeps past the intake port. For simplicity, the ports are shown at the periphery of the housing and induction is depicted as going directly into the chamber. In actual F&S engines the induction tract is more convoluted; the incoming charge goes through the hollow rotor and helps to cool it. Although the charge is heated in the process, and volumetric efficiency suffers, the rotor requires some method of positive cooling. Larger Wankel engines employ oil-cooled rotors, but the complexity of shaft seals, pump, and heat exchanger are prohibitive for small engines.

The compression phase begins as the rotor closes the intake port. The fuel mixture is trapped between the apex seals located at the rotor tips and the housing. As the rotor nears the narrow (or minor) axis of the housing, the spark plug fires. This rotor position is sometimes described as "top dead center" in an agonized attempt to make a comparison between Wankel and conventional engine technology. Dual spark plugs are shown in the illustration, but this approach is rare in small engines. The rationale for it is that the elongated firing chamber imposes long flame-front travel. The second plug, trailing the first by about 10° of rotor rotation, insures more complete combustion, improved economy, and lowered hydrocarbon emissions.

The expansion phase generates torque—that is the purpose of the whole exercise. Superheated air and exhaust gases press against the flank of the rotor and turn it against the fixed reaction gear. It is important to remember that this gear is pinned to the housing and does not move. Power leaves via the main shaft eccentric. Because of the rapidly increasing volume of the working chamber during expansion, there is very little absolute rise in pressure. These engines may be thought of as constant-pressure devices, in contrast to the piston engine, which experiences sudden pressure rise during the expansion phase. In this respect, a Wankel is similar to a big-bore low-speed diesel.

Each rotor face goes through the four cycles in sequence. Events occur simultaneously. The piston engine segregates the working cycles chronologically—the Wankel segregrates them spatially. The 4-cycle piston engine develops one power pulse per each two revolutions of the crankshaft. Since the output shaft of the Wankel is geared for a 3:1 speed advantage over the rotor, and since the rotor is subject to one power pulse per lobe, there is a power pulse during each revolution of the shaft. The Wankel enjoys good volumetric efficiency since there is more time available during induction. The piston engine "inducts" during a nominal 180° of crankshaft rotation, while the Wankel "inducts" over 270° of shaft-output movement.

Expansion involves the same rotational period, viewed in terms of the output shaft. Thus, during a complete revolution of the shaft (360°) there is a period of 90° with no power output. This corresponds to 540° of no torque in a single-cylinder 4-cycle.

In terms of torque output, a single-rotor Wankel is twice as smooth as a four-cylinder 4-cycle, which has a period of 180° with no output.

Perhaps the most significant distinction between piston and rotary engines is that the rotary is characterized by unidirectional movement. In contrast, a piston must come to a dead stop and reverse direction twice during each crankshaft rotation. As pointed out in an earlier chapter, the mass of the piston, wristpin, and half of the connecting rod are subject to reciprocating motions, and are responsible for most of the vibration associated with conventional engines.

The Wankel can be almost perfectly balanced. Two-rotor designs are assembled with the rotors 60° out of phase, thus canceling the forces generated by their gyrations around the reaction gear. Single-rotor designs are not as smooth, but at least counterweight the main shaft eccentric. The KM 3 balance weight is located below the housing, directly above the blade.

CONSTRUCTION

As you can appreciate from Fig. 10-2A, the Wankel engine consists of two moving parts which are related in a rather complex manner. The lobes, or working surfaces of the rotor, form the inner envelope. The outer envelope is defined by the kidney-shaped rotor housing. The clearances between the inner and outer envelope are known as the working chambers, and undergo a constant change in capacity as the rotor gyrates. The outer envelope consists of a generated curve, i.e., one drawn by the relationship between two circles of different diameters. Technically, it is known as a two-node epitrochoidal curve. The rotor is located by the fixed reaction gear which mates with an internal gear on the inner surface of the rotor. The 3:2 ratio between the two is fixed by very complex geometric requirements. The output shaft eccentric is the same diameter as the reaction gear and drives the shaft three times rotor speed. Some explanation is in order.

Suppose the reaction gear was not pinned to the side housing, but was free to rotate. It would turn 1½ times for each rotor revolution (the ratio is 3:2). Since the eccentric is the same diameter as the gear, it could be expected to have the same output ratio. However, the reaction gear is fixed. Instead, the rotor turns on it, and climbs as it turns. The rotor's axis moves in a circle. This circle amounts to a phantom gear which doubles shaft speed over the expected figure.

The rotor in the KM 3 is made of cast iron, as it is in almost all rotary engines (Fig. 10-3). Cast iron is heavy and has poor heat conductivity; it would appear to be a poor candidate for this application. However, aluminum (used in Curtiss-Wright engines and others) requires that the apex seal grooves be reinforced with iron to prevent hammering. An additional insert is required at the hub to accept *reaction* and *eccentric* gear loads. The apex seals (sometimes called tip, or radial seals) are made of high-density cast iron originally developed for piston rings. The working surface of the aluminum rotor housing is protected by the *Elnisil* process. Nickel is flashed on the trochroidal track in conjunction with particles of silicon carbide suspended in an electrolyte bath. Silicon is so much harder than iron that the almost negligible wear incurred is concentrated on the inexpensive seals. The appearance of chatter marks or wave-like depressions, localized in the minor axis of the housing, was an early problem shared by F&S, Toyo Kogyo, and most other Wankel developers.

The researcher's frustration can be realized to a small degree from the Japanese term for these chatter marks—"the scratches of the devil." Elnisil has at least solved the problem handily for F&S.

The KM 3 is unique in that it has only three side seals. The lower seals are eliminated by virtue of a small annular surface which is in rubbing contact with the lower side plate. Seals are not hand-fitted (as was the case in rotary Mazda automobile engines through the 1973 model year), but can be replaced out of inventory stock. The upper side seals are inexpensive stampings, which is another first in rotary engine design. Surface hardening gives adequate wear resistance to the side plates.

Item No	Part No	Description
1	2787 029 000	shroud
2	0242 024 102	shroud nut (4)
3	0240 056 002	shroud bolt (4)
4	3642 001 000	mainshaft nut
5	0518 107 010	lock-washer
6	2766 011 000	fan
7	2740 078 101	thru bolts (10)
8	0246 021 000	washers (10)
9	2711 062 000 / 2787 028 000	side housing (magneto side)
10	2749 024 000	nut (10)
11	0240 057 101	carburetor mounting stud (2)
12	2730 013 000	seal, upper
13	2732 038 000	main bearing, upper
14	0945 122 000	lockring
15	2747 020 000	spacer
16	2732 037 100	eccentric bearing
17	2786 037 005	rotor
18	2715 024 000	button of end seal (3)
19	2739 022 000	spacer
20	2715 025 000	side seal (3)
21	2739 021 000	side seal spring (3)
22	2715 012 000	apex seal spring (3)
23	2715 023 000	apex seal (3)
24	0246 001 001	woodruff key
25	2723 020 000 / 2740 090 001	main or output shaft
26	02786 045 000	rotor housing
27	2734 009 000	reaction gear
28	0232 014 000	locating pin
29	2711 061 000	side housing (pto side)
30	2740 084 000	bolts (3)
31	0932 074 003	main bearing (lower)
32	0945 000 000	lockring
33	2745 009 000	lock ring
34	2730 012 000	seal (lower)
35	2718 014 000	balance weight
36	2740 080 000	set screw

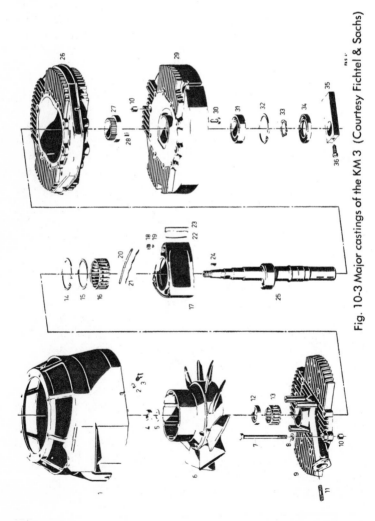

Fig. 10-3 Major castings of the KM 3 (Courtesy Fichtel & Sachs)

DISASSEMBLY AND OVERHAUL

Figure 10-4 illustrates the special tools for this engine. In addition, the factory offers a bench stand which makes servicing somewhat more convenient, although such a stand is hardly necessary. The tachometer, however, is a necessity; it is extremely difficult to estimate the speed of a rotary engine because of the almost total absence of vibration. Another mandatory tool is the fan puller, which may be purchased or made up in the shop using suitable (6 × 60) metric bolts. Ordinary puller tips can be ground to fit the oil seal lips.

Disassembly commences with the removal of the recoil starter, gasoline tank, screen, shroud, carburetor, and muffler. Using the puller, lift the fan. Remove the spring clip (Fig. 10-5, No. 1), woodruff key (4), primary connection (3), and the dust cover (2). Remove the three capscrews securing the stator plate to the end cover.

Lift the stator plate clear, and loosen the ten through-bolts. Tap the upper side housing with a soft-faced mallet to separate it from the main rotor housing. The rotor will be exposed as shown in Fig. 10-6. Next, remove the external balance weight from below the engine and, if necessary, the lower main bearing. This bearing is secured by two snaprings (32 and 33 in Fig. 10-3), and by an interference fit to the lower side housing. Heat the casting to approximately 300°F and drive the bearing down and out. Figure 10-7 gives the dimensions for a bearing mandrel. (This tool is not available from F&S.) The reaction gear is secured by three capscrews (No. 30, Fig. 10-3) which are accessible after the bearing has been removed. Inspect the gear teeth for wear, chipping, and fatigue cracks. In general, it is wise to replace the rotor and reaction gear as a unit.

Seal inspection is the most critical part of overhauling rotary engines. Begin by inspecting the apex grooves for wear. Remove all carbon residue from the grooves and measure the distance between a new seal and the sides of each groove (Fig. 10-8). Permissible clearance is only 0.07 mm (0.0027 in.). Replace the rotor and all seals if the clearance is excessive.

Apex seal height is measured at three points—a, b, and c, in Fig. 10-9A—with the aid of a micrometer. The original dimension is 7.4 mm ± 0.2 mm. Wear must not exceed 0.4 mm along the length of the seal. Also, all three measurements must be within 0.2 mm of each other. The new width is 3.02 mm. Maximum allowable wear is 0.02 mm. The end seal dimensions are shown in the middle drawing (Fig. 10-9B). The as-new

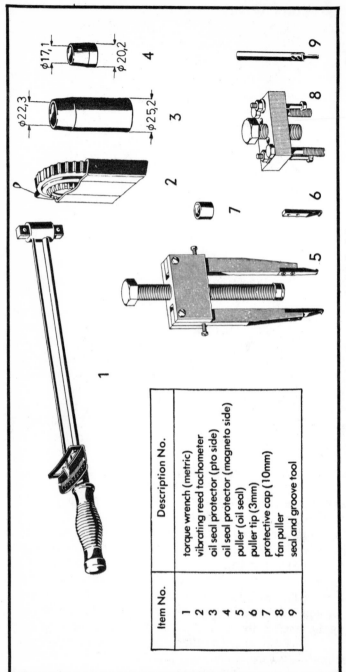

Item No.	Description No.
1	torque wrench (metric)
2	vibrating reed tachometer
3	oil seal protector (pto side)
4	oil seal protector (magneto side)
5	puller (oil seal)
6	puller tip (3mm)
7	protective cap (10mm)
8	fan puller
9	seal and groove tool

Fig. 10-4 Factory tools (Courtesy Fichtel & Sachs)

Fig. 10-5 Magneto assembly (Courtesy Fichtel & Sachs)

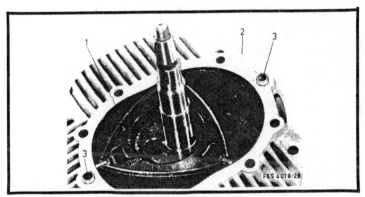

Fig. 10-6 Upper side housing removed showing rotor (1), main rotor housing (2), and locating bushings (3). (Courtesy Fichtel & Sachs)

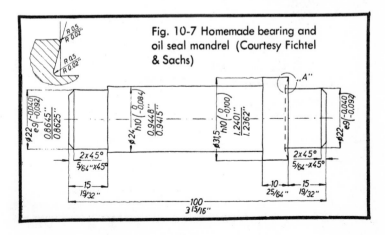

Fig. 10-7 Homemade bearing and oil seal mandrel (Courtesy Fichtel & Sachs)

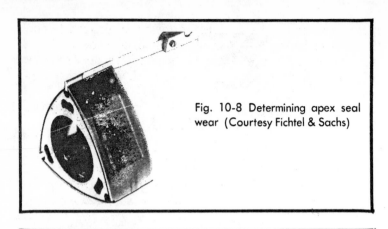

Fig. 10-8 Determining apex seal wear (Courtesy Fichtel & Sachs)

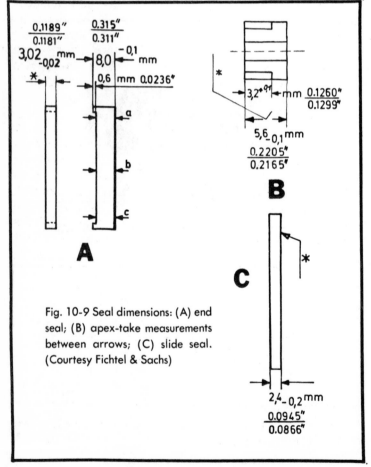

Fig. 10-9 Seal dimensions: (A) end seal; (B) apex-take measurements between arrows; (C) slide seal. (Courtesy Fichtel & Sachs)

height is 5.6 mm which provides for a wear cushion of 0.03 mm. The side seals (Fig. 10-9C) should be replaced when they have worn 0.02 mm. End and side seals should outlast apex seals by a factor of three or four based, on experience with other F&S engines.

Install new springs during every teardown. The spring ends bear against the seals (Fig. 10-10).

The eccentric bearing (Fig. 10-3, No. 16) is located in the rotor cavity by a snapring and spacer. The bearing should not be disturbed unless it is obviously worn or discolored.

Inspect the epitrochoidal curve for deep scratches, abrasions, and for flaking of the Elnisil coating. If the coating fails, the most likely spots will be the minor axis indentations. Of course, if this should happen, the casting must be replaced along with the seals. The side castings are subject to less wear, but should be viewed with suspicion if their friction surfaces show evidence of galling or are compromised by deep scratches. Failure of these parts can be traced to insufficient lubrication or to a dirty air filter. KM 3 side housings cannot be resurfaced but must be replaced with new parts.

ASSEMBLY

Clean all parts in solvent and dry with compressed air. As you assemble, oil the parts liberally with clean oil. Do not mix the oil with any additives. If the main bearings have been removed, warm the castings and press them into place with the tool described in Fig. 10-7. The numbered ends of the bearings are toward the tool. Affix the snaprings and insert the seals. Use the seal protectors shown in Fig. 10-4 to prevent

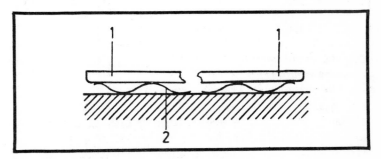

Fig. 10-10 Correct placement of the spring and seal. The spring ends contact the seal tips to give maximum support (Courtesy Fichtel & Sachs)

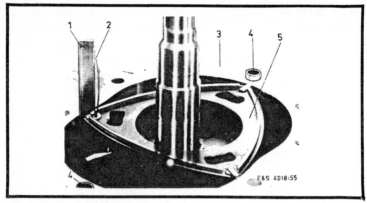

Fig. 10-11 Assembly sequence of output shaft, power-takeoff side housing, rotor housing, guide bushings (4), rotor, seals, and springs (Courtesy Fichtel & Sachs)

damage to the seal lips. You can, if factory tools are not available, wrap the output shaft with a layer of masking tape. Grease the seal lips with Shell Alvania 3 or the equivalent.

Tighten the balance-weight setscrew to 5½ ft-lb. and place the rotor housing over the shaft (No. 3 in Fig. 11). Insert the bushings (4) and slip the rotor on the eccentric. Install the apex, side, and end seals, observing the correct lay of the springs. Lubricate the seals, epitrochoidal track, eccentric bearing, and gear train. Mount the upper side plate and secure with the ten 6×80 capscrews. Tighten in even increments to 6.51 ft-lb.

The air gap between the coil armature and the flywheel is a function of the distance between the output shaft cam and the pole shoe. Measuring from the concentric side of the cam (Fig. 10-12), this distance should be 52.4 mm (2.063 in.). Adjustments are made with the three mounting screws. Next, install the contact points and adjust the gap to 0.016 in.; replace the dust cover and spring clip.

Mount the fan and shroud. Tighten the fan-to-output-shaft nut to 32 ft-lb and assemble the carburetor, muffler, grass screen, and fuel tank. With the engine secured to the mower or to a sturdy bench mount, start the engine and monitor it with a tachometer. Idle—the term is almost a misnomer with this near-constant speed device—is between 2200 and 2700 rpm. Speed without load, at wide-open throttle, is 3100–3200 rpm. Do not attempt to estimate these speeds by sound, vibration, or any other subjective input. Use an accurate tachometer; a rotary engine is as smooth at 6000 rpm as it is at 2000.

If the engine does not perform within these speed limits, remove the tank and mount the rewind starter over the shroud as shown in Fig. 10-12. Note the nuts (1) used as spacers between the starter cover and the shroud. Mount an auxiliary tank above the carburetor as shown. The steel nut (2) secures the operating lever, which in turn positions the throttle. Loosen the nut a turn or so and, with the lever at the full-throttle setting, adjust rpm to specifications. Low-speed adjustments are made with the operating lever in the idle position. Two screws thread into the spark-plug side of the carburetor body. The screw closest to the engine is the low-speed adjustment; the other controls low-speed mixture. The latter should be approximately one-half turn out from finger-tight.

ENGINE ACCESSORIES

The novelty of the rotary engine is confined to the torque-producing parts. It does not extend to the magneto, carburetor, rewind starter, and other accessories.

Magneto

A Ducati magneto, designed for under-flywheel operation, is used on the KM 3 with a provision for remote grounding. Timing is fixed at the factory. The air gap is adjusted as detailed above in the section on engine assembly. Set the contact gap to 0.016 in. with the movable arm on the high

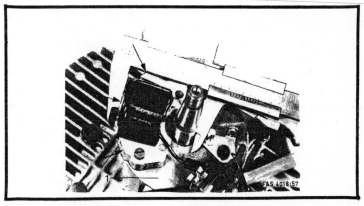

Fig. 10-12 The distance between the shaft and pole piece should be 52.4 mm (Courtesy Fichtel & Sachs)

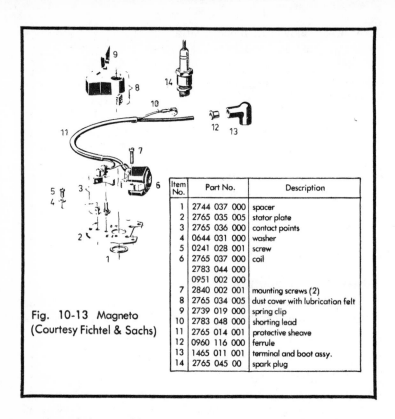

Item No.	Part No.	Description
1	2744 037 000	spacer
2	2765 035 005	stator plate
3	2765 036 000	contact points
4	0644 031 000	washer
5	0241 028 001	screw
6	2765 037 000	coil
	2783 044 000	
	0951 002 000	
7	2840 002 001	mounting screws (2)
8	2765 034 005	dust cover with lubrication felt
9	2739 019 000	spring clip
10	2783 048 000	shorting lead
11	2765 014 001	protective sheave
12	0960 116 000	ferrule
13	1465 011 001	terminal and boot assy.
14	2765 045 00	spark plug

Fig. 10-13 Magneto (Courtesy Fichtel & Sachs)

portion of the cam. Lubricate the wick with several drops of engine oil. The spark plug, a Bosch W 175 T 7, is gaped to 0.020 in. as closely as possible. This plug is a special cold type developed for rotary engines. The company has not authorized substitutions with other brands.

Carburetor

The carburetor is a Bing concentric float model of the type used on many European utility engines. It has a fixed main jet and a pilot (or air-control) needle for low-speed mixture control. Both the choke and throttle are conventional butterfly valves (Fig. 10-14). An air-vane governor is used to compensate for load variations. High-speed adjustment is by way of the governor control lever while idle is a function of the idle-speed screw and the pilot screw. Initial setting on the pilot screw is one-half turn out from seated.

Fuel-line fittings for the tank and carburetor must be coated with a gasoline-proof sealant before assembly. The air

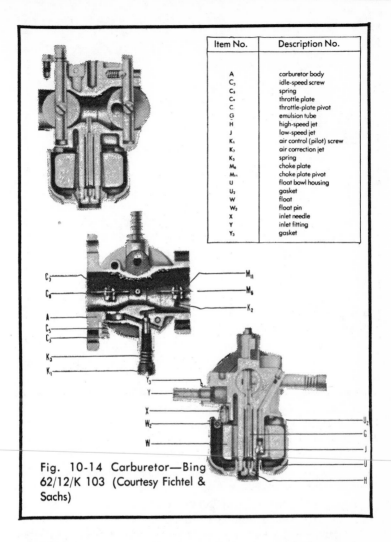

Item No.	Description No.
A	carburetor body
C₃	idle-speed screw
C₄	spring
C₄	throttle plate
C.	throttle-plate pivot
G	emulsion tube
H	high-speed jet
J	low-speed jet
K₁	air control (pilot) screw
K₂	air correction jet
K₃	spring
M₆	choke plate
M₁₁	choke plate pivot
U	float bowl housing
U₂	gasket
W	float
W₂	float pin
X	inlet needle
Y	inlet fitting
Y₃	gasket

Fig. 10-14 Carburetor—Bing 62/12/K 103 (Courtesy Fichtel & Sachs)

filter element is granulated coconut charcoal and should be periodically washed in a solvent bath. Dry with compressed air. (No filter oiling is necessary.)

The rewind starter is similar in design to those discussed in an earlier chapter. When disassembling the main spring take sensible precautions: *Wear leather gloves and safety glasses.* These springs do not have the stored energy of those which fit to ratchet starters, but the free end of a rapidly unwinding spring *can inflict injury.* Preload is one-half to one turn past full rope retraction.

Chapter 11

The Electrics

Electric mowers have never been popular with the public, primarily because of the inconvenience, and because of the potential danger of entangling the power cord. Improvements in storage battery technology have made it possible for OMC, Wheelhorse, and Sears to manufacture models without cords, however. Black & Decker produces a double-insulated motor, and uses other safety features such as the almost universal blade brake. These improvements have minimized the shock hazard associated with ac machines.

The attractions of electric power are numerous. Although some of the twin-bladed geared models howl like banshees, most of the electrics are quiet. The cordless Lawnboy produces a noise level of less than 60 dB at 50 ft—a level that corresponds to the sound intensity of hushed conversation; the hum of electrics is never as grating to the ear as the sound of internal combustion.

Another advantage, at least so far, is found in the fact that electric mowers are very nearly zero-pollution devices. The only emission is the ozone generated by brush sparking; and, at that, in minute amounts. In contrast, a gasoline mower produces clouds of hydrocarbons, quantities of carbon monoxide, and enough oxides of nitrogen to be more than obtrusive. The first of these emissions—hydrocarbons—are released even when the mower isn't running; it will just sit there, adding its share to the smog with gasoline evaporating from its tank. Gasoline evaporates from the tank and fuel bowl.

Economy is another plus for electric mowers. The initial cost is not low, but operating costs are quite negligible. Sears says that the energy cost is on the order of 5 cents per hour. The General Electric ER 8-36 riding mower was a cordless, 8 hp machine with a charger that drew 3.0 kW/hr. This figure, coupled with battery capacity and the known efficiency of oil-fired turboelectric generators, means that during one recharge cycle the mower used less than a quart of fuel oil. One charge allowed the machine to cover up to 10,000 square feet. A gasoline engine of equivalent horsepower would use almost twice the chemical energy, and require lubrication as well.

Gasoline engines require periodic attention. Those with conventional magnetos must have contact points and condensers changed at fairly frequent intervals; spark plugs are also sacrificial items. In contrast, an electric mower requires almost no maintenance. Most feature sealed bearings which are factory-lubricated for the life of the machine. The only parts that have a significant rate of wear are the brushes. The batteries in electric machines (one or more of the lead—acid type) must be charged, have their cases and terminals cleaned, and have their electrolyte level occasionally corrected. As bothersome as this service might seem, it is no more than that demanded by gasoline mowers with electric starters.

Durability is fantastic. Experience has demonstrated that mower decks rusted or cracked from fatigue before anything went wrong with the motor or related parts. Quality control on the cheaper electrics is nothing to write home about, but even these machines should outlast their gasoline counterparts. Electric power is smooth—the only out-of-balance forces generated are due only to manufacturing inaccuracies in the armature and blade assembly.

There are, however, disadvantages to electric mowers. Those with power cords are somewhat hazardous, although statistics are not availabte to show the accident potential of electricity in comparison to gasoline power. The danger of gasoline involves its use both in the machine and in storage. Double insulation at the motor, insulated handles and cord rods, and automatic blade brakes help, but one should realize that a shock potential does exist.

Household current can be lethal. Battery-pack (cordless) machines are safer because of lower voltage. Their shock effect, if any, is more like internal electromassage rather than electrocution. However, there are those who can't tolerate even this milder form of electrical stimulation for medical reasons. To be on the safe side, in any case, don't wear rings, bracelets, wristwatches, or other metallic ornaments when servicing these machines. Some more words of caution: The low-voltage, high-current power supply can give a nasty burn. Lead—acid batteries present other hazards. The electrolyte is a dilute form of sulfuric acid and will attack the skin and destroy clothes. But the major danger is from explosion. Batteries produce hydrogen gas in normal operation, and fairly large amounts are released near the end of the charging cycle. A battery should be treated with all the respect given to a gasoline tank; turn the charger off before breaking the connections on either side.

Another difficulty is in the matter of repair. Gasoline mowers can be repaired, in one fashion or another, by mechanically inclined teenagers, auto mechanics, and by trained small-engine mechanics; also, parts are widely available. In recent years, tuneup parts have been stocked by NAPA and other auto parts jobbers. Electric mowers require less downtime, but when something does go wrong, it may be difficult to find someone to do the work. Most gas engine mechanics have an aversion to electrical machinery and are only too happy to direct the electric mower owner toward the shop door. Even if a willing workman çan be found, parts are few and far between. Several months may elapse between an order to the factory and its delivery. Most electric mowers—Black & Decker, OMC, Sears, Sunbeam, and Wheelhorse—are intended to be serviced at factory depots. Presumably, parts are on hand, although at this writing, service manuals do not exist for some of these machines.

The best advice for the individual owner, or for the shop that wishes to branch out into electrical work, is this: understand the circuit and be willing to chase parts. In the face of a particularly frustrating repair, one might be tempted to substitute parts. I advise against it. And, when it comes to safety-related items—**I say beware!** A three-position switch is a three-position switch—but if it is on a lawnmower handle, it had better be waterproof!

AC MOWER OPERATION

Ac mowers are sold without extension cords; this is comparable to selling rowboats without oars. The dealer should have the correct three-wire extension for his inventory of mowers. It is difficult to generalize about wire gage here, but since most of these mowers draw 6.5—7.5 amps, 16-gage

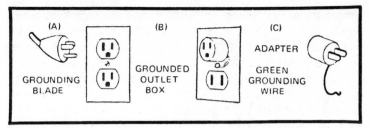

Fig. 11-1 Grounding methods (Courtesy Sunbeam Outdoor Company, a division of Sunbeam Corp.)

should be considered the minimum. Specify a heavier cord if you have a larger motor, or if you are running more than 100 ft from the outlet. However well the mower is insulated, you are asking for a jolt if you do not ground the extension cord at the box. Boxes with grounding blades are best (Fig. 11-1A); they may be grounded during installation to conform with building codes and underwriter's standards. Adapters are available for use with two-prong receptacles. The pigtail wire must be connected to the faceplate screw and the *receptacle must be grounded.*

Automatic reels have gone, by the way, although they once were a feature of some of the better electric mowers. To prevent snarls, support the cord as shown in Fig. 11-2, making generous loops on either side of your hand. Begin near the outlet and follow the simple to-and-fro pattern indicated in Fig. 11-3. The extension cord should always lie on the cut portion of the lawn. Most modern machines feature a cord control rod to keep the cord clear of the machine. At the end of each cut, pivot the mower on its back wheels and flip the rod (shown in dotted lines in Fig. 11-4). The purpose of this technique is to keep the cord on the cut side of the lawn; the same method can be used on machines without a control rod. An alternate suggested by Sunbeam is shown in Fig. 11-5. In this case, the

Fig. 11-2 Wrapping the cord (Courtesy Sunbeam Outdoor Company, a division of Sunbeam Corp.)

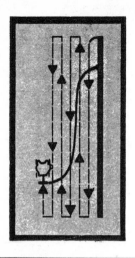

Fig. 11-3 Mowing pattern (Courtesy Black & Decker Company)

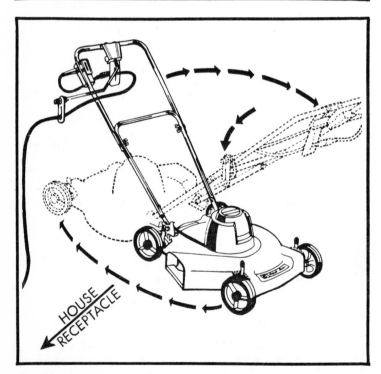

Fig. 11-4 Pivot action at the end of each swath (Courtesy Black & Decker Company)

Fig. 11-5 Alternate cutting pattern (Courtesy Sunbeam Outdoor Company, a division of Sunbeam Corp.)

wheels stay firmly on the ground while the mower is pivoted. The guiding principle—keeping the cord on the cut side of the lawn—is the same.

Should the cord be cut or otherwise damaged, replace it with a UL outdoor cord of the same specifications. *Do not attempt a repair.*

Electric mower safety requires that you observe all the precautions associated with any power mower. Avoid wet grass (potential shock hazard and poor traction), wear sturdy shoes and long trousers, and see that the area is clear of debris. A rotary should be moved across slopes—the ride-on mower should be steered up and down to prevent the machine from turning turtle. Solid objects can be thrown out from under the machine with great force. Discourage spectators and, of course, do not allow children to operate this, or any other, power tool. Best results are obtained when the mower is allowed to operate at its own rate. Forcing the motor will result in a jagged cut and could cause failure from overheating in spite of the circuit protectors.

Adjustments—wheels, grass catcher, mulch plate—should be made with the switch off and the line cord disconnected. Do not leave the machine running and unattended. And before disconnecting, switch the motor off to prevent a surprise start when it is hooked up again.

CORDLESS MOWER OPERATION

After some experimentation with massive batteries, the industry has become standardized to a 36V power supply consisting of three 12V batteries in series (Fig. 11-6). Battery capacity and current drain determine range. Riding mowers which may have as many as three motors, naturally employ heavier batteries than do rotary push-type mowers.

The batteries are shipped dry; you have to fill the cells with the packaged electrolyte to the prescribed level—above the plates but below the filler cap. Some batteries have factory-sealed vent tubes. If yours is one of these, cut the melded ends of the tubes and route them down through the battery-pack case; the tubes should exit from the case, but should not be pinched or crimped. A clogged vent can result in a battery explosion as gas pressure builds up in the cell near the end of the charge cycle. Put the battery pack on light charge to bring the cells up to a specific gravity of 1.1275 on a temperature-corrected hydrometer. In order to avoid acid spillage and as an added safety precaution, do not tilt the machine with the batteries in place. Remove the battery pack or the individual batteries from their mounts.

The charger may be carried on board or may be a separate fixture (Fig. 11-7). Some chargers are entirely automatic and have an indicator lamp whose brightness gives

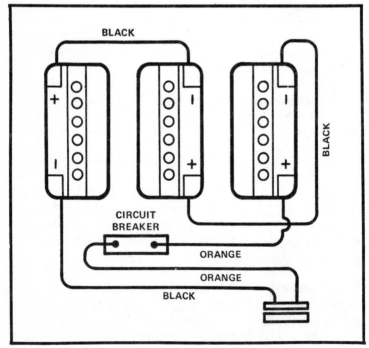

Fig. 11-6 Battery connections (Courtesy OMC)

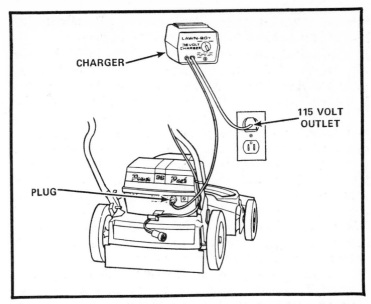

Fig. 11-7 A fully automatic charger. Disconnect from the wall before making or breaking the motor connection (Courtesy OMC)

some clue to the rate of charge. These devices cut off automatically when battery voltage equals charger input. Other chargers employ a timer. Timers allow a better control of the state of charge than can be had from a fully automatic unit; but their use does require some discretion. Overcharging is not the healthiest condition in the world for a battery, but is better than chronic undercharge. The amount of charge required depends upon the condition of the batteries, the temperature of the electrolyte, and current drain since the last charge. New batteries require some break-in and will require more than normal charge current. As the battery ages it will require longer periods of charge. Experiment by moving the timer dial one increment back at a time. Charge and discharge the batteries at least twice before deciding that the duration of charge is insufficient.

As the thermometer drops, chemical activity slows, and the battery will require longer periods on the charger. For example, if one were engaged in mowing a lawn at 30°F, the recharge period would be twice normal. On the other hand, do not attempt to charge a battery when the temperature climbs above 110°F; the electrolyte will overheat and damage the

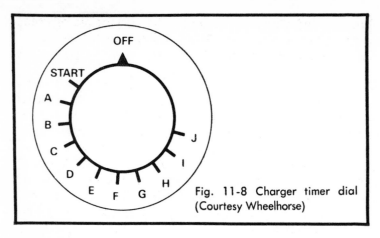

Fig. 11-8 Charger timer dial
(Courtesy Wheelhorse)

plates. As a general rule, electrolyte temperature must not be allowed to climb beyond 125°F.

Deep discharge is bad for storage batteries, especially when they are new. For best life, try to keep the batteries above half-discharge; this corresponds to a specific gravity of 1.200—1.220 on a temperature-corrected hydrometer. However, in many applications this will not be possible. OMC offers additional battery packs which can be kept hot and installed as needed. Another technique is to quick-charge the batteries for 10 minutes or so during the mowing session.

When using a battery-powered machine, attack the heaviest grass and the most difficult terrain first; save the light work and trimming until last when the batteries are weak. Insofar as possible, cut in a regular pattern as shown in Fig. 11-9. Try to avoid the power-robbing effect of cutting over

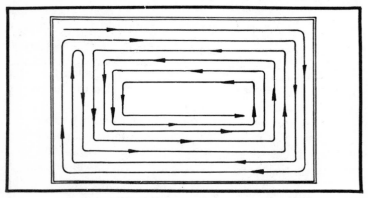

Fig. 11-9 Mowing pattern (Courtesy Wheelhorse)

clippings, and allow as small an overlap as possible between swaths. Mowing wet grass, which collects under the deck, clogs the discharge stream and encourages rust.

CLEANING

Do not splash water on the motor, deck, or switch assembly. The mower should be wiped free of dust with a damp rag. Grass and dirt upset the air stream and can reduce mower efficiency markedly. With the extension cord or batteries disconnected, clean the underside of the deck with a putty knife. The extension cord should be disconnected from the outlet and wiped off with a damp rag. Do not use petroleum-based solvents on the cord since these solvents will attack the insulation.

LUBRICATION

Most motors feature sealed bearings which are lubricated for life; Black & Decker machines have an oil fitting for the upper shaft bearing which should be lightly oiled with SAE 20 motor oil, seasonally. Do not overoil. Machines that use bronze bushings at the wheels (the majority) require periodic lubrication to reduce wear and to prevent the shaft from rusting between uses. Others are fitted with nylon bushings and do not take kindly to lubrication. Riding mowers normally have grease fittings at the wheels and underdeck bearings. Lubricate them with chassis grease, using a *hand*-powered gun—air-powered grease guns can blow bearing seals. Other lubrication points are drive chains and steering linkages.

BLADES

All currently produced electric mowers use rotary blades mounted singly or in tandem. The latter arrangement (Fig. 11-10) has advantages in terms of space utilization, reduction

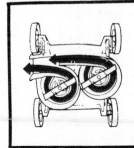

Fig. 11-10 Tandem blades (Courtesy Sunbeam Outdoor Company, a division of Sunbeam Corp.)

of blade tip speeds and (when belt-driven) in protecting the motor for blade impacts. Inspect the blades for warpage as discussed previously, and sharpen as needed. Light sharpening can be done with the blades mounted if the switch is off and the extension cord is disconnected at the motor (or the power pack removed). Turn the machine on it side, lock the blade with a length of board; if you prefer, use a drill-mounted grinder as shown (Fig. 11-11)—and wear safety glasses. Drive belts should be replaced in matched pairs. Black & Decker sells a belt kit for their machines that includes a tension gage to enable you to make the correct adjustment.

MOTOR REPAIR

Lawnmower motors, whether ac or dc, share many common design features and can be lumped together accordingly for this discussion. Field coils are wired in series parallel with the armature, although there is a recent tendency to use permanent magnets for the fields. This allows reduced current consumption, a saving in weight, and better reliability than the electromagnetic variety. Another characteristic of these motors and allied circuitry is the liberal use of diodes—for rectification and to simplify the switches. Most motors feature brush-fed armatures and a mechanical brake to stop the blade when power is off (compression does this on gasoline-powered machines).

Unlike the situation of a few years ago when mowers were usually powered by GE or Westinghouse appliance motors, the

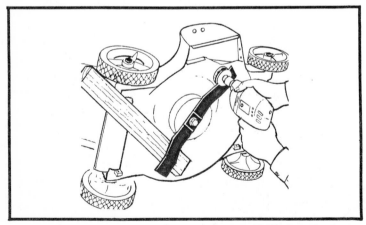

Fig. 11-11 Blade sharpening (Courtesy Black & Decker)

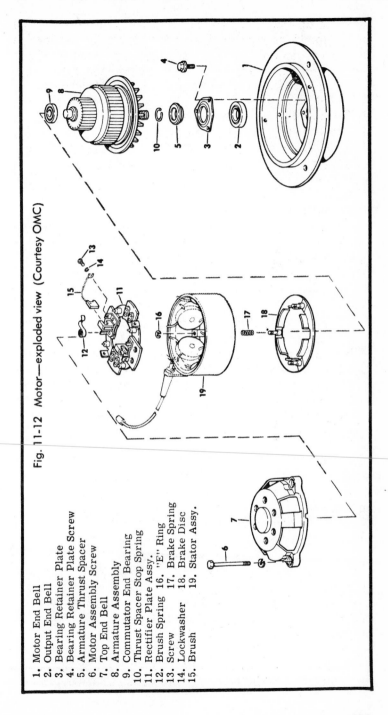

Fig. 11-12 Motor—exploded view (Courtesy OMC)

1. Motor End Bell
2. Output End Bell
3. Bearing Retainer Plate
4. Bearing Retainer Plate Screw
5. Armature Thrust Spacer
6. Motor Assembly Screw
7. Top End Bell
8. Armature Assembly
9. Commutator End Bearing
10. Thrust Spacer Stop Spring
11. Rectifier Plate Assy.
12. Brush Spring
13. Screw
14. Lockwasher
15. Brush
16. "E" Ring
17. Brake Spring
18. Brake Disc
19. Stator Assy.

current crop employs single-purpose motors, often designed and built by the firm that makes the machine. Consequently, there is no interchangeability of parts between brands.

Figure 11-12 is the reference picture for our discussion. It is an exploded view of an OMC ac motor with an integral brake and cooling fan (the line fields operate on dc.)

To disassemble, disconnect the extension cord and the motor lead (part 19). Remove the blade and the mounting bolts which secure the lower bell (1) to the deck. Remove the capscrews (6) and lift the top end bell (7) from the stator assembly. At this point the brushes are accessible for inspection or replacement.

Most problems with these motors originate at the brushes, commutator, or brush holders. The brushes should not be worn to less than half their original length; and they should move freely in the holder. Remove carbon dust and gum with a good nontoxic solvent (such as an aerosol TV tuner cleaner), and wipe clean with paper towels or a lintless rag. Check the brush holders for damage or looseness at the mounting rivets; repair as necessary. Make a continuity check between the grounded brush holder and the frame with an ohmmeter, and make the same check on the insulated holder.

Lift the armature out of the stator assembly (19 in Fig. 11-12) and place it in a lathe or on accurately ground V-blocks to determine if the shaft is bent. How much bend can be tolerated is one of those specifications that hasn't been published yet for electric mowers. New shafts seem to be true to within 0.001 in.; much more than this will cause vibration severe enough to be hazardous to the operator. In most instances, attempts to straighten the motor shaft have proved impractical.

If the commutator is out-of-round, make the correction explained in a previous chapter.

If the bearings are noisy or sloppy, replace them with *equivalents*—don't "make do." The motor depicted is typical in that the lower bearing must be pulled and installed with a press. You can use a wide-capacity vise, if you support the motor at the top end of the shaft and make a suitable collar for the bearing; a large socket can do yeoman duty here.

Electrical checks for the stator have been outlined previously in the chapter on starter motors. Procedures are the same. One should be careful in making leakage tests as they require the use of line current. Lightly turn the commutator as necessary, and undercut the insulation.

The series and shunt field coils can be tested for grounds and for continuity with a low-voltage ohmmeter or a test lamp. Field coil resistance measurements (which would indicate the presence of internal loop-to-loop shorting) would be meaningless since "optimum" figures are not available—when in doubt, use new parts.

The ferrous brake disc (18) is pulled upward by the field coils when the motor is energized. With the switch off, the springs (17) force the disc down on the brake assembly. The brake assembly should bring the blade to a smooth halt in 10 seconds or less.

The diodes for field coil rectification are located in the external circuit and may be checked with an ohmmeter in the usual fashion.

TROUBLESHOOTING THE MOTOR

Troubleshooting these motors is not difficult. We will begin with the worst failure—the motor does not run and the fuse blows as soon as the switch is pulled. This can only mean a short circuit. Check the line cord with an ohmmeter; with the cord disconnected at both ends there should be infinite resistance between the two prongs. Examine the insulation for breaks and abrasions which would allow current to go to ground. Do the same for the lead connecting motor-to-control box. The diodes and the switch should also be checked for grounding.

If these elements are good, the problem could only be in the motor proper. In most cases, the armature will have given up the ghost; it will be found to have internal shorts or be grounded to the frame. A faulty series field coil can also be responsible, although field coil failure is fairly rare.

The next case—the fuse blows when the motor runs—almost always involves internal motor circuitry. The armature may be shorted (in which case the commutator will be spark-eroded and discolored) as well as might be either field winding; or the shunt winding may be open.

Failure to run without fuse problems is symptomatic of an open circuit. Check the circuit from the line cord to the diodes, through the series field, to the brushes (poor contact is often the difficulty). Replace worn brushes and retest.

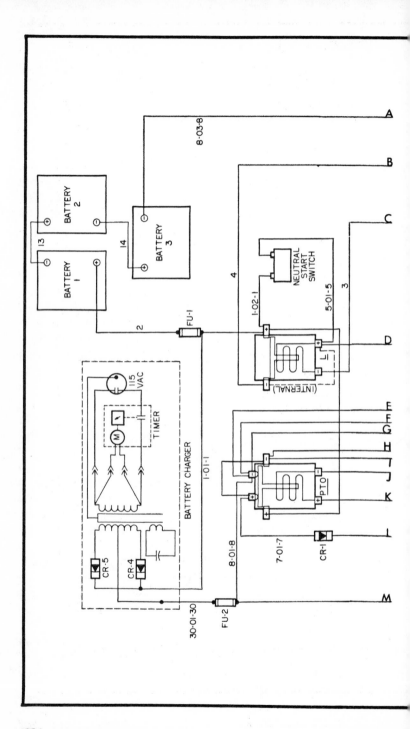

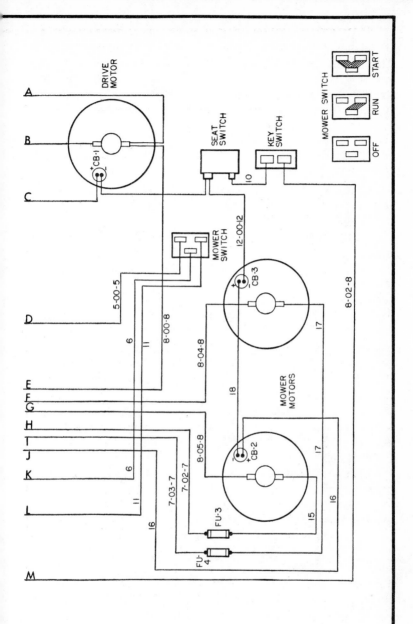

Fig. 11-13 A-65 wiring diagram (Courtesy Wheelhorse)

397

High (4000 rpm plus) no-load speeds, coupled with a lack of power, point to a shorted series winding, or to an open shunt. The shunt winding acts as a kind of governor for series—parallel motors. No-load speeds under 3000 rpm can involve the armature and diodes or can be caused by mechanical drag—as from a sticking brake or flattened bearing.

CONTROL CIRCUITS

Control circuits for ac machines are quite simple, consisting of a switch and (possibly) single or paired diodes. Resistance tests through the components are adequate for diagnosis.

Dc machines have more elaborate circuitry. Figure 11-13 illustrates the circuit layout for the Wheelhorse (see General Electric) riding mower. This machine has a full-wave rectifier in the permanently mounted charger, three 12V batteries in series, and three motors. Two of these motors drive the blades, while the third one propels the machine through a multispeed transmission. The main switch is labeled MOWER SWITCH and is shown to the right center of the diagram. It has three positions: OFF, RUN, and START. OFF kills all circuitry except that associated with the battery charger. The key switch has two positions. OFF kills all circuits except that of the battery charger. ON will energize the drive motor, assuming (1) the seat switch is depressed, (2) the key switch is on, (3) the neutral start switch is closed and, on some models, (4) the clutch is depressed.

There is evidence of careful design throughout this machine. For example, three automotive-type fuses are employed to protect the charger and the individual mower motors. In addition, each motor has its own circuit breaker. Normally, the drive motor breaker does not open—unless you try to push down trees—but the mower motors *will* stop if the machine is forced too rapidly into heavy grass. When this happens, turn the mower switch off, wait a few seconds, and restart.

Charger problems can be evidenced by always-dead batteries, although the sure way to ascertain the cause is to connect a voltmeter across the output (it should read close to 40V) or an ammeter in series with one of the leads. The rate of charge depends upon the condition of the batteries, electrolyte

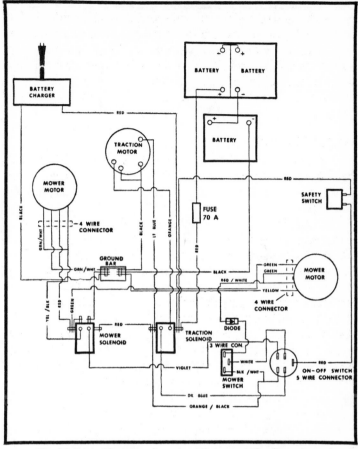

Fig. 11-14 A-60 wiring diagram (Courtesy Wheelhorse)

temperature, and the like, but these chargers should be able to deliver a healthy 20A. Charger failure can usually be traced to the fuse, labeled FU-2 in the diagram. Should either of the diodes (CR-4 and CR-5) open, charging current (and dc output voltage) will be cut drastically.

The various switches can be checked with a voltmeter or ohmmeter. If you elect to jump the switches, be careful—these are safety related items. For example, jumping the neutral switch could set the machine in motion.

The A-60 diagram is shown in Fig. 11-14. Although the layout is similar to the A-65, there are minor differences.

The OMC circuit for the Model 5801 rotary is quite simple. It consists of a 3-position switch, central circuit breaker, and

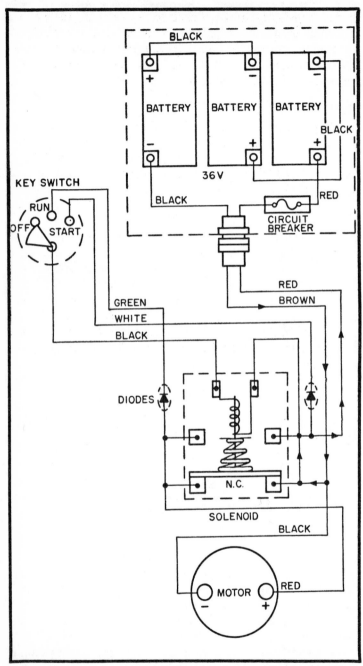

Fig. 11-15 Lawnboy Model 5801 wiring diagram (Courtesy OMC)

diodes. The diodes simplify the circuit by reducing the number of switch positions. On START, the solenoid is energized. In terms of the diagram in Fig. 11-15, the solenoid element moves upward, bridging the second set of contacts. Upon release, the spring-loaded switch returns to the RUN position.

TROUBLESHOOTING CONTROL CIRCUITS

Begin with the motor. Wire directly from a charged battery pack. Should the motor fail to run, or run sluggishly, disassemble and make the checks previously described. If, on the other hand, the motor operates normally, you can be sure the problem lies in the control circuitry.

Check all connections before turning to the individual control circuit components. Begin with the handlebar switch—disconnect all leads. There should be infinite resistance between the RUN and START contacts and zero resistance between the rotating switch element and each of the two contacts as they are brought into the circuit. Next, check the diodes. They should pass current in one direction and effectively block it in the other. The solenoid is the most fragile component of the group. Make a resistance test between the moving element and the stationary contacts, or simply jump the solenoid.

Glossary

AC—Alternating current, or current that reverses its direction at regular intervals.

Air bleed—A passage in the carburetor through which air can seep or bleed into fuel moving through a fuel passae.

Air cleaner—A device, mounted on the carburetor or connected to the carburetor, through which air must pass before entering the carburetor air horn. A filtering device in the air cleaner removes dust and dirt particles from the air.

Air-cooled engine—An engine cooled by air circulating between cylinders and around cylinder head as opposed to the liquid-cooled engine by a liquid passing through jackets surrounding the cylinders.

Air filter—A filter through which air passes, and which removes dust and dirt particles from the air. Air filters are placed in passages through which air must pass, as in a crankcase breather, air cleaner, etc.

Air horn—That part of the air passage in the carburetor which is on the atmospheric side of the venturi. The choke valve is located in the air horn.

Ampere—Unit of electric-current-flow measurement. The current that will flow through a 1-ohm resistance when 1 volt is impressed across the resistance.

Antifriction bearing—A bearing of the type that supports the imposed load on rolling surfaces (balls, rollers, needles), minimizing friction.

Antiknock—Refers to substances that are added to automotive fuel to decrease the tendency to knock when fuel-air mixutre is compressed and ignited in the engine cylinder.

Armature—The rotating assembly in a direct-current generator or motor. Also, the iron piece in certain electrical apparatus that completes a magnetic (and in many cases, an electric) circuit.

Atmospheric pressure—The weight of the atmosphere per unit area.

Atom—The smallest particle, or part, of an element, composed of electrons and protons and also of neutrons (with exception of hydrogen).

Atomization—The spraying of a liquid through a nozzle o that the liquid is broken into tiny globules or particles.

Axial—In a direction parallel to the axis. Axial movement is movement parallel to the axis.

Axis—A center line. The line about which something rotates or about which something is evenly divided

Axle—A cross support on a vehicle on which supporting wheel, or wheels, turn. There are two general types: live axles that also transmit power to the wheels and dead axles that transmit no power.

Backfiring—Preexplosion of fuel-air mixture so that explosion passes back around the opened intake valve and flashes back through the intake manifold.

Backlash—The backward rotation of driven gear that is permitted by clearance between meshing teeth of two gears.

Baffle—A plate or shield to divert the flow of liquid or gas

Ball bearing—A type of bearing which contains steel balls that roll between inner and outer races.

Battery—A device consisting of two or more cells for converting chemical energy into electrical energy.

Battery charging—The process of supplying a battery with a flow of electric current to produce chemical actions in the battery; these actions reactivate the chemicals in the battery so they can again produce electrical energy.

BDC—Bottom dead center; the position of the piston when it reaches the lower limit of travel in the cylinder.

Bearing—A part in which a journal pivots, a pin turns, or revolves. A part on or in which another part slides.

Bendix drive—A type of drive used in a starter which provides automatic coupling with the engine flywheel for cranking and automatic uncoupling when the engine starts.

Bevel gear—One of a pair of meshing gears whose working surfaces are inclined to the center lines of the driving and driven shafts.

Block—*See* Cylinder block.

Blow-by—Leakage of the compressed fuel-air mixture or burned gases from combustion, passing piston and rings and into the crankcase.

Blower—A mechanical device for compressing and delivering air to engine at higher than atmospheric pressure.

Bore—The diameter of engine cylinder hole. Also diameter of any hole; as, for example, the hole into which a bushing is fitted.

Boss—An extension or strengthened section, such as the projections within a piston which supports the piston pin.

Brake band—A flexible band, usually of metal with an inner lining of brake fabric, which is tightened on a drum to slow or stop drum rotation.

Brakedrum—Metal drum mounted on car wheel or other rotating member; brake shoes or brake band, mechanically forced against it, causes it to slow or stop.

Brakeshoes—The curved metal part, faced with brake lining, which is forced against the brake drum to produce braking or retarding action.

Brake horsepower—The power actually delivered by the engine which is available for driving the vehicle.

Brake lining—A special woven fabric material with which brake shoes or brake bands are lined; it withstands high temperatures and pressures.

Bronze—An alloy consisting essentially of copper and tin.

Brushes—The carbon or carbon and metal parts in a motor or generator that contact the rotating armature commutator or rings.

Bushing—A sleeve placed in a bore to serve as a bearing surface.

Bypass—A separate passage which permits a liquid. gas. or electric current to take a path other than that normally used.

Cam—A moving part of an irregular form designed to move or alter the motion of another part.

Capacitance—That property of a circuit which tends to increase the amount of current flowing in a circuit for a given voltage or to delete in its entirety.

Capacitor (condenser)—A device for inserting the property of capacitance into a circuit: two or more conductors separated by a dielectric.

Carburetor—The device in a fuel system which mixes fuel and air and delivers the combustible mixture to the intake manifold.

Cell—A combination of electrodes and electrolyte which converts chemical energy into electrical energy. Two or more cells connected together form a battery.

Centrifugal advance—The mechanism in an ignition distributor by which the spark is advanced or retarded as the engine speed varies.

Centrifugal force—The force acting on a rotating body. which tends to move its parts outward and away from the center of rotation.

Charging rate—The rate of flow. in amperes. of electric current flowing through a battery while it is being charged.

Choke—A device in the carburetor that chokes off. or reduces. the flow of air into the intake manifold: this produces a partial vacuum in the intake manifold and a consequent richer fuel-air mixture.

Circuit—A closed path or combination of paths through which passage of the medium (electric current. air. liquid. etc.) is possible.

Clockwise—Direction of movement. usually rotary. which is the same as movement of hands on the face of a clock.

Clutch—The mechanism located in the power train. that connects the engine to. or disconnects the engine from. the remainder of the power train.

Coil—In electrical circuits. turns of wire. usually on a core and enclosed in a case. through which electric current passes.

Coil spring—A type of spring made of an elastic metal such as steel. formed into a wire or bar and wound into a coil.

Combustion—A chemical action. or burning: in an engine. the burning of a fuel-air mixture in the combustion chamber.

Combustion chamber—The space at the top of the cylinder and in the head in which combustion of the fuel-air mixture takes place.

Commutation—The process of converting alternating current which flows in the armature windings of direct current generators into direct current.

Commutator—That part of rotating machinery which makes electrical contact with the brushes and connects the armature windings with the external circuit.

Compression—Act of pressing into a smaller space or reducing in size or volume by pressure.

Compression ratio—The ratio between the volume in the cylinder with the piston at bottom dead center and with the piston at top dead center.

Compression rings—The upper rings on a piston: the rings designed to hold the compression in the cylinder and prevent blow-by.

Compression stroke—The piston stroke from bottom dead center to top dead center during which both valves are closed and the gases in the cylinder are compressed.

Concentric—Having a common center. as circles or spheres. one within the other.

Condenser—*See* Capacitor.

Conductor—A material through which electricity will readily flow.

Connecting rod—Linkage between the crankshaft and piston. usually attached to the piston by a piston pin and to the crank journal on the crankshaft by a split bearing and bearing cap.

Cooling fins—Thin metal projections on air-cooled-engine cylinder and head which greatly increases the heat-radiating surfaces and helps provide cooling of engine cylinder.

Core—An iron mass. generally the central portion of a coil or electromagnet or armature around which the wire is coiled.

Counterclockwise—Direction of movement. usually rotary. which is opposite in direction to movement of hands on the face of a clock.

Crankcase—The lower part of the engine in which the crankshaft rotates. In automotive practice. the upper part is lower section of cylinder block while lower section is the oil pan.

Crankcase breather—The opening or tube that allows air to enter the crankcase and thus permit crankcase ventilation.

Crankcase dilution—Dilution of the lubricating oil in the oil pan by liquid gasoline seeping down the cylinder walls past the piston rings.

Crankcase ventilation—The circulation of air through the crankcase which removes water and other vapors. thereby preventing the formation of water slude and other unwanted substances.

Crankshaft—The main rotating member or shaft of the engine. with cranks to which the connecting rods are attached.

Current regulator—A magnetic-controlled relay by which the field circuit of the generator is made and broken very rapidly to secure even current output from the generator and prevent generator overload from excessive output. (One of the three units comprising a generator regulator.)

Cutout relay—An automatic magnetic switch attached to the generator to cut out generator circuit and prevent overcharging of the battery. *See* Circuit breaker.

Cycle—A series of events with a start and finish. during which a definite train of events takes place. In the engine. the four piston strokes (or two piston strokes on 2-stroke cycle engine) that complete the working process and produce power.

Cylinder—A tubular-shaped structure. In the engine. the tubular opening in which the piston moves up and down.

Cylinder block—That part of an engine to which, and in which. other engine parts and accessories are attached or assembled.

Cylinder head—The part of the engine that encloses the cylinder bores. Contains water jackets (on liquid-cooled engine) and valves (on I-head engines).

DC—Direct current, or current that flows in one direction only.

Damper—A device for reducing the motion or oscillations of moving parts, air, or liquid.

Dead axle—An axle that simply supports and does not turn or deliver power to the wheel or rotating member.

Deceleration—The process of slowing down. Opposite to acceleration

Detonation—In the engine, excessively rapid burning of the compressed fuel-air mixture so that knocking results.

Diaphragm—A flexible membrane, usually made of fabric and rubber in automotive components, clamped at the edges and usually spring-loaded; used in fuel pump, vacuum pump, distributor, etc.

Differential—A mechanism between axles that permits one axle to turn at a different speed than the other and, at the same time, transmits power from the driving shaft to the axles.

Disc brake—A type of brake which depends upon contact between two or more discs for its effect. One or more of the discs may be faced with brake lining.

Dual-ratio axles—Axles which contain a mechanism for changing driving ratio of the wheels to either high or low ratio. Two-speed differential.

Dynamometer—A device for measuring power output of an engine.

Eccentric—Off center

Efficiency—Ratio between the effect produced and the power expended to produce the effect.

Electricity—A form of energy that involves the movement of electrons from one place to another, or the gathering of electrons in one area.

Electrode—Either terminal of an electric source; either conductor by which the current enters and leaves an electrolyte.

Electrolyte—The liquid in a battery or other electrochemical device, in which the conduction of electricity is accompanied by chemical decomposition.

Electromagnet—Temporary magnet constructed by winding a number of turns of insulated wire into a coil or around an iron core; it is energized by a flow of electric current through the coil.

Electron—Negative charged particle that is a basic constituent of matter and electricity. Movement of electrons is an electric current.

Energy—The capacity for performing work.

Engine—An assembly that burns fuel to produce power, sometimes referred to as the power plant.

Evaporation—The action that takes place when a liquid changes to a vapor or gas.

Exhaust stroke—The piston stroke from bottom dead center to top dead center during which the exhaust valve is opened so that burned gases are forced from the engine cylinder.

Exhaust valve—The valve which opens to allow the burned gases to escape from the cylinder during the exhaust stroke.

Field—In a generator or electric motor the area in which a magnetic flow occurs.

Field coil—A coil of wire, wound around an iron core, which produces the magnetic field in a generator or motor when current passes through it.

Field frame—The frame in a generator or motor into which the field coils are assembled.

Filter—A device through which gas or liquid is passed; dirt, dust, and other impurities are removed by the separating action.

Float—In the carburetor, the metal shell that is suspended by the fuel in the float bowl and controls a needle valve that regulates the fuel level in the bowl.

Flywheel—The rotating metal wheel, attached to the crankshaft, that helps level out the power surges from the power strokes and also serves as part of the clutch and engine-cranking system.

Foot-pound—A unit of work done in raising 1 pound avoirdupois against the force of gravity to the height of 1 foot.

Force—The action that one body may exert upon another to change its motion or shape.

Four-stroke-cycle engine—An engine that requires four piston strokes (intake, compression, power, exhaust) to make the complete cycle of events in the engine cylinder.

Fuel pump—The mechanism in the fuel system that transfers fuel from the fuel tank to the carburetor.

Fuse—A circuit-protecting device which makes use of a substance that has a low melting point. The substance melts if an overload occurs, thus protecting other devices in the system.

Gasket—A flat strip, usually of cork or metal, or both, placed between two surfaces to provide a tight seal between them.

Gasoline—A hydrocarbon, obtained from petroleum, is suitable as an internal combustion engine fuel.

Gear ratio—The relative speeds at which two gears turn; the proportional rate of rotation.

Gears—Mechanical devices to transmit power or turning effort from one shaft to another; more specifically, gears which contain teeth that engage or mesh upon turning.

Generator—In the electrical system, the device that changes mechanical energy to electical energy for lighting lights, charging the battery, etc.

Generator regulator—In the electrical system, the unit which is composed of the current regulator voltage regulator, and circuit breaker relay.

Ground—Connection of an electrical unit to the engine or frame to return the current to its source.

Gusset plate—A plate at the joint of a frame structure of steel to strengthen the joint.

Helical—In the shape of a helix, which is the shape of a screw thread or coil spring.

High-speed circuit—In the carburetor, the passages through which fuel flows when the throttle valve is fully opened.

High tension—Another term for high voltage. In the electrical system, refers to the ignition secondary circuit since this circuit produces high-voltage surges to cause sparking at the spark plugs.

Hydrometer—A device to determine the specific gravity of a liquid. This indicates the freezing point of the coolant in a cooling system or, as another example, the state of charge of a battery.

Idle—Engine speed when throttle is fully released; generally assumed to mean when engine is doing no work.

Idle circuit—The circuit in the carburetor through which fuel is fed when the engine is idling.

Idle gear—A gear placed between a driving and a driven gear to make them rotate in the same direction. It does not affect the gear ratio.

Idling adjustment—Adjustment made on the carburetor to alter the fuel-air mixture ratio or engine speed on idle.

Ignition—The action of setting fire to; in the engine, the initiating of the combustion process in the engine cylinders.

Ignition advance—Refers to the spark advance produces by the distributor in accordance with engine speed and intake manifold vacuum.

Ignition coil—That component of the ignition system that acts as a transformer and steps up battery voltage to many thousand volts; the high voltage then produces a spark at the spark-plug gap.

Ignition switch—The switch in the ignition system that can be operated to open or close the ignition primary circuit.

Ignition timing—Refers to the timing of the spark at the spark plug as related to the piston position in the engine cylinder.

I-head—A type of engine with valves in the cylinder head.

In-line engine—An engine in which all engine cylinders are in a single row or line.

Insert—A form of screw thread insert to be placed in a tapped hole into which a screw or bolt will be screwed. The insert protects the part into which the hole was tapped, preventing enlargement due to repeated removal and replacement of the bolt.

Insulation—Substance that stops movement of electricity (electrical insulation) or heat (heat insulation).

Intake manifold—That component of the engine which provides a series of passages from the carburetor to the engine cylinders through which fuel-air mixture can flow.

Intake stroke—The piston stroke from top dead center to bottom dead center during which the intake valve is open and the cylinder receives a charge of fuel-air mixture.

Integral—Whole; entire; lacking nothing of completeness.

Interference—In radio, any signal received that overrides or prevents normal reception of the desired signal. In mechanical practice, anything that causes mismating of parts so they cannot be normally assembled.

Intake valve—The valve in the engine which is opened during the intake stroke to permit the entrance of fuel-air mixture into the cylinder.

Internal combustion engine—An engine in which the fuel is burned inside the engine, as opposed to an external combustion engine where the fuel is burned outside the engine, such as a steam engine.

Jackshaft—An intermediate driving shaft.

Jet—A metered opening in an air or fuel passage to control the flow of fuel or air.

Journal—That part of a shaft that rotates in a bearing.

Knock—In the engine, a rapping or hammering noise resulting from excessively rapid burning or detonation of the compressed fuel-air mixutre.

Knuckle—A joint or parts carrying a hinge pin which permit one part to swing about or move in relation to another.

Laminated—Made up of thin sheets, leaves, or plates.

Lean mixture—A fuel-air mixture that has a high proportion of air and a low proportion of fuel.

Lever—A rigid bar beam of any shape capable of turning about one point, called the fulcrum; used for transmitting or changing force or motion.

Leverage—The mechanical advantage obtained by use of lever; also an arrangement or combination of levers.

L-head—A type of engine with valves in the cylinder block.

Lubrication—The process of supplying a coating of oil between moving surfaces to prevent actual contact between them. The oil film permits relative movement with little frictional resistance.

Magnet—Any body that has the ability to attract iron.

Magnetic field—The space around a magnet which the magnetic lines of force permeate.

Magnetic flux—The total amount of magnetic induction across or through a given surface.

Magnetic pole—Focus of magnetic lines of force entering or emanating from magnet.

Magnetism—The property exhibited by certain substances and produced by electron (or electric current) motion which results in the attraction of iron.

Magneto—A device that generates voltage surges, transforms them to high-voltage surges, and distributes them to the engine cylinder spark plugs.

Main bearing—In the engine, the bearings that support the crankshaft.

Member—Any essential part of a machine or structure.

Meshing—The mating or engaging of the teeth of two gears.

Molecule—The smallest particle into which a chemical compound can be divided.

Motor—A device for converting electrical energy into mechanical energy.

Muffler—In the exhaust system, a device through which the exhaust gases must pass; in the muffler, the exhaust sounds are greatly reduced.

Mutual induction—Induction associated with more than one circuit, as two coils, one of which induces current in the other as the current in the first changes.

Negative terminal—The terminal from which electrons depart when a circuit is completed from this terminal to the positive terminal of generator or battery.

Needle valve—Type of valve with rod-shaped, needle-pointed valve body which works into a valve seat so shaped that the needle point fits into it and closes the passage; the needle valve in the carburetor float circuit is an example.

North pole—The pole of a magnet from which the lines of force are assumed to emanate.

Nozzle—An orifice or opening in a carburetor through which fuel feeds into the passing air stream on its way to the intake manifold.

Octane rating—A measure of the antiknock value of engine fuel.

Ohm—A measure of electrical resistance. A conductor of one ohm resistance will allow a flow of one ampere of current when one volt is imposed on it.

Ohmmeter—A device for measuring ohms resistance of a circuit or electrical machine.

Oil control rings—The lower rings on the piston which are designed to prevent excessive amounts of oil from working up into the combustion chamber.

Oil pan—The lower part of the crankcase in which a reservoir of oil is maintained.

Oil pump—The pump that transfers oil from the oil pan to the various moving parts in the engine that require lubrication.

Overhead valve—Valve mounted in head above combustion chamber. Valve in I-head engine.

Parallel circuit—The electrical circuit formed when two or more electrical devices have like terminals connected together (positive to positive and negative to negative) so that each may operate independently of the other.

Period—The time required for the completion of one cycle.

Permanent mgnet—Piece of steel or alloy in which molecules are so aligned that the piece continues to exhibit magnetism without application of external influence.

Piston—In an engine, the cylindrical part that moves up and down in the cylinder.

Piston displacement—The volume displaced by the piston as it moves from the bottom to the top of the cylinder in one complete stroke.

Piston pin—The cylindrical or tubular metal pin that attaches the piston to the connecting rod (also called wrist pin).

Power—The rate of doing work.

Power stroke—The piston stroke from top dead center to bottom dead center during which the fuel-air mixture burns and forces the piston down so the engine produces power.

Power-take-off—An attachment for connecting the engine to power driven auxiliary machinery when its use is required.

Preignition—Premature ignition of the fuel-air mixture being compressed in the cylinder on the compression stroke.

Proton—Basic particle of matter having a positive electrical charge, normally associated with the nucleus of the atom.

Radial—Pertaining to the radius of a circle.

Rpm—Revolutions per minute, a measure of rotational speed.

Rectifier—An electrical device that changes alternating current to direct current.

Relay—In the electrical system, a device that opens or closes a second circuit in response to voltage or amperage changes in a controlling circuit.

Resistor—In an electrical system, a device made of resistance wire, carbon, or other resisting material, which has a definite value of resistance and serves a definite purpose in the system by virtue of that resistance.

Rheostat—A resistor for regulating the current by means of variable resistance and serves a definite purpose in the system by virtue of that resistance.

Rich mixture—Fuel-air mixture with a high proportion of fuel.

Ring gear—A gear in the form of a ring such as the ring gear on a flywheel or differential.

Rod cap—The lower detachable part of the connecting rod which can be taken off by removing bolts or nuts so the rod can be detached from the crankshaft.

Roller bearing—A type of bearing with rollers positioned between two races.

Rotor—A part that revolves in a stationary part; especially the rotating member of an electrical mechanism.

SAE—Society of Automotive Engineers.

SAE horsepower—A measurement based upon number of cylinders and cylinder diameter.

Separator—In the storage battery, the wood, rubber, or glass mat strip used as insulator to hold the battery plates apart.

Series circuit—The electrical circuit formed when two or more electrical devices have unlike terminals connected together (positive to negative) so that the same current must flow through all.

Shim—A strip of copper or similar material, used under a bearing cap for example, to adjust bearing clearance.

Short circuit—In electrical circuits, an abnormal connection that permits current to take a short path or circuit, thus bypassing important parts of the normal circuit.

Solenoid—A coil of wire that exhibits magnetic properties when electric current passes through it.

South pole—The pole of the magnet into which it is assumed the magentic lines of force pass.

Specific gravity—The ratio of the weight of a substance to weight of an equal volume of chemically pure water at 39.2°F.

Spline—Slot or groove cut in a shaft or bore; a splined shaft onto which a hub, wheel, etc., with matching splines in its bore is assembled so the two must engage and turn together.

Spray unit—A form of overrunning clutch; power can be transmitted through it in one direction but not in the other.

Storage battery—A lead-acid electrochemical device that changes chemical energy into electric energy. The action is reversible; electric energy supplied to the battery stores chemical energy.

Stroke—The movement, or the distance of the movement, in either direction, of the piston travel in an engine.

Sulfation—A crystalline formation of lead sulfate on storage battery plates.

Taper—To make gradually smaller toward one end; a gradual reduction in size in a given direction.

TDC—Top dead center; the position of the piston when it reaches the upper limit of travel in the cylinder.

Tension—A stress caused by a pulling force.

Throttle—A mechanism in the fuel system that permits the driver to vary the amount of fuel-air mixture entering the engine and thus control the engine speed.

Throttle valve plate—The disk in the lower part of the carburetor air horn that can be tilted to pass more or less fuel-air mixture to the engine.

Thrust—A force tending to push a body out of alignment. A force exerted endwise through a member upon another member.

Tie rod—A rod connection in the steering system between wheels.

Timing—Refers to ignition or valve timing and pertains to the relation between the actions of the ignition or valve mechanism and piston position in the cylinder.

Torque—A twisting or turning effort. Torque is the product, of force times the distance, from the center of rotation at which it is exerted.

Torque rod—Arm or rod used to insure accurate alignment of an axle with the frame and to relieve springs of driving and braking stresses.

Torque wrench—A special wrench with a dial that indicates the amount of torque being applied to a bolt or nut.

Universal joint—A device that transmits power through an angle.

Vacuum—A space entirely devoid of matter.

Vacuum advance—The mechanism on an ignition distributor that advances the spark in accordance with vacuum in the intake manifold.

Valve seat—The surface, normally curved, against which the valve operating face comes to rest, to provide a seal against leakage of liquid, gas, or vapor.

Valve spring—The compression-type spring that closes the valve when the valve-operating cam assumes a closed-valve position.

Valve tappet—The part that rides on the valve-operating cam and transmits motion form the cam to the valve stem or push rod.

Valve timing—Refers to the timing of valve closing and opening in relation to piston position in the cylinder.

Valve train—The train of moving parts to the valve that causes valve movement.

Vapor lock—A condition in the fuel system in which gasoline has vaporized. as in the fuel line. so that fuel delivery to the carburetor is blocked or retarded.

Venturi—In the carburetor. the restriction in the air horn that produces the vacuum responsible for the movement of fuel into the passing air stream.

Volatility—A measurement of the ease with which a liquid turns to vapor.

Volt—Unit of potential. potential difference. or electrical pressure.

Voltage regulator—A device used in connection with generator to keep the voltage constant and to prevent it from exceeding a predetermined maximum. (One of the three units comprising a generator regulator.)

Volumetric efficiency—Ratio between the amount of fuel-air mixture that actually enters an engine cylinder and the amount that could enter under ideal conditions.

Wobble plate—That part of a special type of pump (wobble pump) which drives plungers back and forth as it rotates to produce pumping action. It is a disc. or plate. set at an angle on a rotating shaft.

Work—The result of a force acting against opposition to produce motion. It is measured in terms of the product of the force and the distance it acts.

Worm gear—A gear having concave. helical teeth that mesh with the threads of a worm. Also called a *worm wheel*.

Index

422